A Gardener's Book of Names

Michael Paffard

By the same author:

Inglorious Wordworths Hodder & Stoughton 1973
The Unattended Moment S.C.M. Press 1976
Thinking About English Wardlock 1978

A Gardener's Book of Names

BY

Michael Paffard

New Millennium
310 Kennington Road, London SE11 4LD.

Printed and bound by B.W.D. Ltd. Northolt, Middx.
Issued by New Millennium*
ISBN 1 85845 034 9
*An imprint of The Professional Authors' & Publishers' Association

In memory of
Professor Alan Gemmell
who suggested the writing of this book
as a task for the author's retirement.

Contents

	Page no
Introduction	vii
Chapter 1: A Little Bit About Language	1
Chapter 2: Plants Get Proper Names	9
Chapter 3: Plants Get Common Names	17
Chapter 4: Buddle, Fuchs & Co	33
Chapter 5: An Anthology Of Annuals	45
Chapter 6: A Patchwork Of Perennials	63
Chapter 7: A Pick Of Pot Plants	77
Chapter 8: A Sheaf Of Shrubs	91
Chapter 9: A Clutch Of Climbers	101
Chapter 10: A Grabble Of Ground Cover	109
Chapter 11: A Recumbence Of Rock Plants	117
Chapter 12: A Bushel Of Bulbs	127
Chapter 13: A Wheelbarrow Full Of Weeds	139
Chapter 14: Variegated Vegetables	153
Chapter 15 :A Peck Of Pot Herbs	167
Chapter 16: A Case Of Fruit & Nuts	175

Chapter 17: A Bushful Of Birds	185
Chapter 18: A Bunch Of British Botanists	199
Chapter 19: The Name On The Gate	205
Chapter 20: Miscellaneous Matters	211
Bibliography	229
Index	233

Introduction

In one of his delightful books, Wilfred Blunt tells of a gardener who had a mischievous way of dealing with one of the problems we all encounter: when he was asked by a visitor to his garden for the name of a plant which he could not remember, he would confidently say it was Bishop's Bigamy. This invariably satisfied the enquirer. The moral, I suppose, is that most people take names for granted and are uncurious about what they mean and where they come from.

Names can certainly be a problem, as every gardener knows. Many plants have several common English names, as we shall see in Chapter 2, and many common names are used for several different and quite unrelated plants. This can obviously be a source of confusion which is not entirely solved if we stick to the scientific names. They can be somewhat of a mouthful of unmemorable botanical Latin and Greek: it requires considerable concentration to remember *Acer palmatum 'Dissectum Atropurpurium'* all the way to the garden centre and courage to ask for it confidently, and that is only part of the problem. For reasons we shall explore in Chapter 3, botanists keep changing their minds about the classification of plants and therefore about the proper names they should have. Horticulture is not always quick to adopt the new names approved by the scientific taxonomist: old and new names for a plant may both be current for a considerable period of time and that is another source of confusion. I have tried to include both the old and the new, and place my entry under whichever of the two I think the reader is most likely to come across in visiting gardens or studying growers' catalogues. Both will be in the index.

The index, indeed, is essential in making the best use of the book. After the first three introductory chapters, it has been arranged for dipping and browsing, but the index will enable you to use it as a reference book. When a generic name has been explained in one chapter, it will not reappear elsewhere. *Cornus* is in the chapter on shrubs, so you will look in vain for *Cornus canadensis,* whose name needs no further explanation, under Ground Cover where it might very reasonably be expected. Similarly *Convolvulus* is among the

Weeds rather than the Climbers, and Thymus with the Pot Herbs rather than the Rock Plants.

Where a plant is a native member of the British flora, or has close relatives which are, I have drawn attention to this fact and also chosen illustrative examples in the first three chapters mainly from among our wild plants in the hope that this will enhance the interest of the book for naturalists as well as gardeners. Today many people are becoming aware of the importance of their gardens as reserves for our endangered wildlife and even planting exclusively native species. Trees, ferns and grasses have had to be neglected for reasons of space, as have all but a pick of popular pot plants from among the thousands of tender things which can be grown in a heated greenhouse. Indeed, in every chapter I have had to be selective, including I hope all the commoner things but necessarily excluding many that are not often encountered in cultivation. There are more or less comprehensive dictionaries of plant names listed in the bibliography on p. 229 and I am much indebted to them (even when their expert authors disagree!).

And disagree they certainly do. Etymology is a field full of booby-traps and pitfalls for the unwary: experts cross it with great circumspection and their conclusions are often heavily qualified with ifs and buts. Taxonomy, too, which is the science of classifying living things, is full of problems and uncertainties, and for a similar reason. In both studies one is concerned to find out where things as we know them today - words in one case and organisms in the other - came from, how they evolved. That means enquiring into a time long before written records existed. Bring the two subjects together as I have had to do, and you create a quagmire on which angels might well fear to tread even if the Almighty had equipped them with celestial gumboots. I have tried to avoid guess-work and give accurate information but I may not always have used the three ps, perhaps, possibly, probably, as often as I should have done, nor have I entered into all the shifting arguments between the experts. Strict scholarly caution can make for difficult reading and this book is not for experts but for the curious and alert gardener or naturalist. That is why I thought it worth finding space for the names we give our houses, for some bipeds - gardeners and garden birds - and for an assortment of other legged or legless creatures which the gardener comes across.

The typescript has been read by my good friend and former colleague Dr Keith Goodway and by that lynx-eyed error-detector, my wife: both have made many valuable corrections and suggestions for which I am very grateful. For such errors and infelicities as remain, however, I must accept total responsibility. I cannot hope to have avoided what Dr Johnson calls, in the preface to his great Dictionary, "a few wild blunders, and risible absurdities from which no work of such multiplicity was ever free"; indeed, I shall be well satisfied if they prove to be only a few.

CAROLI LINNÆI

Naturæ Curioſorum *Dioſcoridis Secundi*

SYSTEMA NATURÆ

IN QUO

NATURÆ REGNA TRIA,

SECUNDUM

CLASSES, ORDINES, GENERA, SPECIES,

SYSTEMATICE PROPONUNTUR.

Editio Secunda, Auctior.

STOCKHOLMIÆ

Apud GOTTFR. KIESEWETTER.

1740.

Systema Natura by Carl Linnaeus,
the greatest name of them all. Title page of 1740

Chapter 1

A Little Bit About Language And A Cautionary Tale

Names are words, words are language and the study of language which used to be called 'philology' is now usually called 'linguistics' by scholars. If you are interested in names and where they came from, the branch of linguistics that will be of help is etymology which studies just that, where words came from and how they have changed their forms and sometimes their meanings in the course of time. Let us take an example: what sort of rubbish did a medieval town produce? We hardly need a social historian to tell us that there were no tin cans, plastic, glass bottles, packaging or newspapers. Oddments of wood were burnt on the hearth and everything else went on the midden: it was all bio-degradable as we say today, with one exception: only bones were excluded. They were piled up on some vacant lot, picked clean by scavenging birds and rats and eventually burnt. So the bonfire on which you burn your garden rubbish was originally a bone-fire and the big Oxford Dictionary quotes from the splendidly named printer Wynkyn de Worde writing in 1515 and telling us that on the feast of St John the people of London celebrated with "fyres of clene bones and noo wood and this is called a bone fyre". Quite a stink it made, I imagine.

It should not be supposed, however, that the social historian (or the geographer for that matter) can always be trusted about the original meanings of names which is the proper business of the etymologist. For instance, at one time, readers of the official history of Teddington were told that it is situated at the highest point to which the Thames is tidal and that Teddington therefore means 'tide-ending-town'. Plausible enough you might think, but the early records of the name show that it meant no such thing: it was the 'ton' or home of the people of an Anglo Saxon chief called Tudda. Linguists know of many similar mistakes.

Let us take another cautionary example. I live in a village called Keele. A distinguished professor of Geography at the local university used to tell his students that the word 'keel' is a Norse word (which is true) and that since the Norsemen used to turn their boats upside down and sleep underneath them, 'keel' could once mean a roof-ridge or watershed as well as the central timber of a ship's bottom (which is also true). He then went on to say that the parish of Keele is on the watershed of England since one of the two steams which rise in the parish flows east into the Trent and on to the North Sea while the other reaches the Irish Sea to the west by way of the Mersey. (And that is also true; very good geography in fact). Unfortunately he left the students believing that Keele therefore meant a watershed, whereas the early spellings of the place-name leave little doubt that it was 'ky-hill' or hill of cows long before people would have known that it was a watershed.

The moral of this story is not that professors can be wrong: we all know that I hope. It is rather that, if you want to know what a name means, the name of a plant or a place or a person or whatever, you must do what an etymologist would do: you must study the earliest recorded spellings of that name and bring to them a knowledge of how the English language has changed and developed since. Use an etymological dictionary by all means, but don't necessarily believe what you read elsewhere in print. People took an interest in names and wrote books about them before the study of language was very well understood. Some of those old books have been reprinted recently and are not without interest, but much of what they contain is ingenious guess-work and picturesque supposition.

When we come to look at plant names in later chapters, we shall often find that they contain the same sort of guess-work and mistakes. Many of the English names by which we know both wild and cultivated species were given to them by the old herbalists in the sixteenth and seventeenth centuries. They were collectors of plants who were mainly interested in their useful properties and had no scientific way of classifying them. It is hardly surprising if they sometimes got it wrong when translating from other languages or trying to match the plants they knew with those described by the ancient authorities in Latin or Greek. Gerard, for instance, gave the name Horse-radish to *Armoracia rusticana* from South East Asia because he misunderstood the German *Meerettich*, thinking *Meer* was 'mare' rather than 'sea'. The Germans called it the 'sea-radish' because it was imported from overseas. Actually 'Horse-radish' is quite appropriate for a large, coarse plant.

William Turner knew that Spinach had been introduced as a pot-herb from Spain and so assumed the name meant 'Spanish herb', whereas the Spanish *espinaca* comes from the Arabic *isbanakh.*

Parkinson generally improved on Gerard and Turner, but managed to add his own crop of mistakes as well. Danewort or Daneweed were the old names he recorded for Dwarf Elder (*Sambucus ebulus*), adding by way of explanation, "it grows in great plentie about Slaughterford where there was a great fight with the Danes". However, he was doubly wrong. There are two places called Slaughterford: the one in Gloucestershire means 'muddy-ford' (slough) and the one in Wiltshire means 'ford by the blackthorns' (sloe-thorn-ford); neither has anything whatever to do with bloodshed. And 'danes' or 'dains' was the old English word for diarrhoea: Dwarf Elder or Danewort is a powerful purgative.

These three Tudor herbalists, like all their contemporaries, can only have had the haziest notion of how their English language had come to be the way it was. Parkinson knew about the Danes in England, but not about what they had or had not contributed to the language. In fact they had left behind them a layer in our lexis. Gardeners mean by 'layering' a method of propagation, but when linguists use the word they refer to the way the vocabulary of a language like English has been made up of successive layers resulting from a history of invasion and settlement.

English is, of course, basically Anglish, made up that is of the Germanic dialects spoken by the warlike, pagan tribes of Angles and Saxons who settled here after the collapse of the Roman Empire. They adopted little in the way of language from the bottom two layers of the pile, that is to say from the Celtic language of the Romanized Britons which was the ancestor of modern Welsh (except for a lot of place- and river-names) or from the Latin of their Roman rulers. Rather more Latin crept into Anglo Saxon (or Old English as it is usually called) after St Augustine arrived at Canterbury in AD 597 and restored the Christian religion. Less than two centuries later the Anglo Saxon kingdoms were invaded and settled by Danes and

Norsemen bringing a layer of Scandinavian into the dialects of the language, particularly those spoken to the north and east of Watling Street. By the beginning of the eleventh century Knut, King of Denmark, ruled as well England, Scotland and Norway, and then fifty years later came William the Conqueror and his followers, superimposing a layer of Norman French on the English of the natives. Finally, at the close of the Middle Ages, there was an invasion of another kind: with the rediscovery of classical learning and the invention of printing, English was all but swamped by a flood of learned words from Latin and Greek. They were being referred to disparagingly in Shakespeare's day as "ink-horn terms".

The result of all this layering is that we have a huge vocabulary of about half a million words and can very often choose between two, three or more nearly synonymous names for one thing. So we have 'bloom' (German *Blumen*) and 'blossom' both from before 1066, as well as 'flower' (French *fleur*) from the Normans, and the botanist may use the ink-horn word 'inflorescence' as a technical term. We have 'orchard' from Old English *ortgeard* (a yard or enclosure in which you grew orts or edible things - not just fruit originally) and we have 'garden' from the Norman *jardin*. And so on and so forth. Our multiple plant names often arose in the same way: 'Willow' is Old English, 'Withy' Scandinavian, 'Osier' Norman French, while 'Sallow' is obviously derived from the Latin *Salix*. 'Furze' is Old English, 'Whin' Scandinavian and 'Gorse' possibly Celtic, related to Welsh *gorest*, a waste place. 'Heather' and 'Heath' are Old English, 'Ling' Scandinavian, and so we could go on. These are all native plants with names from different layers of English. When we come to introduced plants, we find that their names have often come with them from a score of different countries all around the globe, from China to Peru, from the Arctic to the Antipodes.

As words come down to us, they get battered by time: they change as the language changes until it would be impossible to guess from their modern form what their original meaning was, and plant names are no exception. Who could guess that to the Anglo Saxons 'lady' was 'loaf-kneader' and 'lord' was 'loaf-guardian'? The plant Lords-and-Ladies (*Arum maculatum*) has other male-and-female names like 'Bulls-and-Cows', 'Mares-and-Stallions', 'Adam-and-Eve', 'Robin-and-Joan'. Its erect spadix sheathed in a spathe has suggested to the pure-minded 'Lady's Finger', 'Bloody Man's Finger', 'Parson-in-the-Pulpit' and, more equivocally, 'Ram's Horn'. But it has had irresistible sexual resemblances for most country people down the ages. 'Cuckoo-Pint' and 'Wake-Robin' are common and typical names: 'Cuckoo-Pint' is a cleaned up version of 'Cuckoo's pintle' or penis; (Parson's or Priest's pintle and Dog's dibble are, or at least were, widespread too); Wake-Robin is from the French *robinet* meaning 'spigot', 'cock' and hence, as in English, also 'penis'.

I choose Lords-and-Ladies as an example for several reasons: it illustrates the fact that the more striking or unusual the appearance of a plant, the larger the number of names it is likely to have acquired; it shows that where an older name might be thought coarse, it has been changed to make it politer, usually in Victorian times when the pursuit of botany was a popular occupation for young ladies; (Linguists call this 'amelioration'); and it illustrates the old belief that a plant's appearance gave a sign or 'signature' to its use: Lords-and-Ladies gave a phallic sign that it would 'stirr up the affections to conjunction' as one old writer puts it and so it was eaten as an aphrodisiac. Geoffrey Grigson says that it was known in London at one time as Portland sago because the tubers were sent there from the Isle of Portland and used as a substitute for arrowroot.

We can illustrate more briefly some of the things that have happened to plant names as they moved, say, from Latin through French into medieval English or from pre-Conquest English or Scandinavian down the centuries to the Tudor herbalists. In the first place, long names are very likely to get shortened: Orpine (*Sedum telephium*, Livelong) is a contraction, by way of French, from Latin *auripigmentum*, 'gold-colouring'. It is a somewhat mysterious name since the flowers are pinkish-purple and it was not, so far as I can discover, used as a dye-plant. Perhaps the name originally belonged to one of the yellow-flowered Sedums. To take an English example, Herb Bennet (*Geum urbanum*, Wood Avens) is a shortening of Herb St Benedictus; its clove-scented root is one of the ingredients of the liqueur Benedictine. More misleadingly, Mayweed (*Anthemis cotula*, sometimes called Stinking Chamomile) which was hated by harvesters because it blisters the skin, has nothing to do with the month of May; it was originally Maydeweed or Maithenweed. The opposite sometimes happens but is much less common than contraction: Meadowsweet (*Filipendula ulmaria*) was originally Meadsweet because it was used for flavouring mead and not because it grew in meadows; it is typically a waterside plant.

Secondly, an unaccented first syllable is likely to drop off, a sort of leprosy of language which linguists call aphesis. We have seen Spanish *espinaca* become Spinach; similarly Old French *eschalotte* becomes Shallot (*Allium ascalonicum* from the town of Ascalon in Palestine); and Latin *expurgare*, to purge, by way of French becomes Spurge. Potato in regional dialects becomes 'tater' or 'tatty' which shows the same thing happening.

Thirdly, we have the insertion of a sound to make a word easier to pronounce (which is called epenthesis). King Knut becoming Canute and the 'p' in surnames like Thompson and Simpson show this happening. Try saying Old English *spinel* quickly several times and you will find a 'd' creeps into it naturally to make Spindle. Similarly *bremel* has become Bramble.

Fourthly, the sounds in the middle of a word sometimes hop around and change places (which is called metathesis). It is particularly likely to affect the letter 'r' which has always caused English speakers more trouble than any other letter. So, Old English *bridd* has become bird, Old French *brunet* (little brown one) has become Burnet and *pimprinelle* Pimpernel. Small children often metathesise when they are attempting to say a big adult word: I remember my daughter aged two or three saying 'tephelone' and 'ephelant' for telephone and elephant.

Finally there is 'folk-etymology' which is what linguists call the ordinary person's attempt to make some sense of words which he finds unintelligible. We have seen Parkinson who did not understand the Old English word *daines* for diarrhoea associating it with the Danes whom he had heard of, and, at a more sophisticated level, the professor explaining the place-name Keele from his knowledge of the geography of the place. Hardy's heroine Tess of the D'Urbervilles was called Durbeyfield: Norman *ville* very often becomes 'field' in English place-names. There is a place in County

Durham called Bear Park but it was originally *Beau-repayre* which didn't make much sense to the local peasants: it was a Norman lord's 'beautiful retreat', and nothing to do with bears, whatever they may tell you up there. This process is particularly likely to happen when people without a classical education try to domesticate the barbarous Latin and Greek of the botanists. In my part of the country, Rhododendrons are called 'Rosy-Dandums' (which is not far wrong, since the Greek means 'rose-tree'). Sparrow-grass for Asparagus is another familiar example. I believe I have even detected an instance of folk etymology in the great Dr Johnson's Dictionary: he defines Codlin as "a hard apple usually eaten coddled" but Codlin was originally '*Coeur-de-lion*', the lion-hearted apple, because of its hardiness and had nothing to do with being eaten coddled.

If you know just a little about the different layers of English from which our plant names often come and the sort of changes that are likely to have overtaken them in the passage of time, you will often get an insight into their original meanings and, thus, how they were regarded and what they were used for by our ancestors. A final anecdote will demonstrate this and round off this anecdotal chapter. In 1940, when I was a schoolboy, we were evacuated to Devonshire and I found a shrubby St John's Wort which I had never seen before growing at the edge of a shady wood. As well as yellow flowers, it had green, red and black berries on it simultaneously and I remember thinking, 'I bet that's wickedly poisonous!' and refraining from touching it. I couldn't have been more wrong. I looked it up at home and found that it was called Tutsan (*Hypericum androsaemum*) but though I was then better informed, I was not much wiser. If I had been told that *androsaemum* meant 'man's blood' from the colour of the juice of the berries, my suspicion of the plant would hardly have been allayed. Had I known that many of our plant names come from the French of the Normans I might, even then, have deduced that 'Tutsan' came from French *toute-saine*, 'all wholesome'. I don't remember when I discovered that its West Country names are Bible-leaf, Book-leaf or Sweet-leaf. When the leaves are green they have no smell, but put one between the pages of a book and in a few days it will be giving off a delightful aroma, somewhat like the scent of phlox, which lasts for years. It is a treasure of a plant; one of my favourites.

Manypeeplia Upsidownia
Edward Lear makes fun of botanical names in his
Nonsense Botany, 1871.

Chapter 2
Plants Get Proper Names

Biologically speaking, we all belong to one species: black and white, wise and stupid, we all have the binomial name of *Homo sapiens*. However, for everyday purposes of identification, we make do with a surname which links us to our relatives and paternal ancestors and one or more first names to distinguish us from the nearer of them. Plant names work in much the same way, though the rules about naming plants are much stricter than any to which parents have to conform. Often our neighbours will refer to us as 'the Smiths' or 'the Joneses' just as we will talk of roses or onions, but sometimes they need to be more specific and so do we. There are hundreds of John Smiths around and confusion is avoided when it matters, to banks or the DSS for example, by giving each of them a unique number. There are hundreds of roses too, (well, about 250 species actually) but each of them has to have a unique and specific name. For all I know, somebody somewhere may be numbering them for the benefit of a botanical computer, but for the time being if we want to avoid confusion about plants we need to know their specific names which come after their generic or surnames, not before as with our own names.

This is not going to be a book - even a simple one - about classifying plants and giving them names, which is a big and complicated science called taxonomy. I would not be qualified to write such a book. Much of the time even good and knowledgeable gardeners get along with the English or common names of plants and it is their history and meanings that are our main concern. However, the common names we use are often not English at all originally: Anemone is from Greek for 'daughter of the wind' and Saxifrage from Latin for 'stone-breaker'. What we are doing when we use these names is taking the scientific name for a family or genus of plants and using it as a common English name. You may talk of Indian Cress or Larks' Heels but I guess you are more likely to call them Nasturtiums from Latin for 'nose-twister' because of their pungent smell; and, scientifically, that is *Tropaeolum majus* because its leaves made Linnaeus think of the shields which the Greeks hung up as battle 'trophies'. It would not, therefore, be practicable, even if it were desirable, to keep common and scientific names entirely separate. To the botanist there are more than 150 species of Anemone and twice as many Saxifrages and even an unambitious gardener may sometimes want names to distinguish between them. If you got *Anemone japonica*, which is two or three feet tall, when you wanted the dwarf *A. blanda* for the front of the rockery, you would not be pleased. So, in practice, we all need scientific as well as English names even if we find some of them an unmemorable mouthful. (One has some sympathy with whoever it was who described botany as the science of insulting beautiful things in Latin and Greek.) One of the secondary objects of this book will be to show that they are less of a mouthful and more memorable if you know a little of what they mean. Some of our 'common' names are from other languages too: Edelweiss from German 'noble-white', Tulip from Turkish for 'turban', Tomato from Aztec and so on, but the Greek and Latin languages are by far the most important sources. More of that later.

One of the problems of classifying and naming plants is the sheer number to be dealt with. Theophrastus is often called 'the father of botany': he was a pupil of Aristotle and died in 285 BC having written a history of plants in many volumes. He had the advantage of studying all the new plants brought back from the campaigns of Alexander the Great but still only described 500 types of plant. (Today we know that number of the Viola family alone.) Two centuries later, another Greek, Dioscorides, a medical man who served with Nero's army, described the uses of 600 plants. (We know that number of different mosses.) Later still, Pliny the Elder wrote about 1000 different plants in his Natural History. (We have twice that number of Euphorbias and twenty times that number of Orchids.) Pliny died in the great eruption of Vesuvius which buried Pompeii in AD 79 and it was to be 1500 years before there were any major advances in botanical knowledge.

Renaissance herbalists like Leonhart Fuchs, after whom Fuchsias are named, almost always start with Pliny and go on to add other plants which they knew. They were mainly interested in collecting and growing things which were useful or ornamental. By the time of James I's royal botanist, Mathias de l'Obel, a Fleming whose name has been given to the Lobelia, about 4000 different plants were recognised. The older herbalists and apothecaries had 'gone simpling', that is to say collecting useful plants from the wild, but by the seventeenth century horticulture as a business was developing. High prices were being paid for rare plants which were thought to have particular curative properties or gave their owners status. Noblemen vied with each other to be first to have new introductions from the Americas or elsewhere in the world. The tulip-mania in Holland was only the most extreme example of a booming trade in plants, new varieties being bred as well as introduced.

Just as this profitable business was developing, a new breed of scientist was beginning to study the plant world with disinterested curiosity and the need for a proper system of naming plants was becoming very evident. Several obstacles stood in the way of progress. There was no mechanism for calling conferences and getting international agreement and, as in astronomy, so in botany, religion impeded science. Although it was recognised that there were 'natural' groups of plants, the umbellifers for example, it was a stake-and-faggot matter to suggest that they might have developed from a common ancestor. The Church taught that God made the species on the third day of Creation and several seventeenth century botanists refrained from publishing their work for fear of being charged with heresy.

Another piece of quasi-religious superstition stood in the way of progress and that was the so-called Doctrine of Signatures. Theophrastus Bombast von Hohenheim, who lived from 1493 to 1540 and called himself Paracelsus, is generally credited with propounding the theory, though the general idea goes back to classical and possibly to prehistoric times. Paracelsus wrote in Latin but an English version of the doctrine can be found in *The Art of Simpling* by William Coles, a contemporary of Culpeper. "Though Sin and Sathan hath plunged Mankinde into an Ocean of Infirmities, yet the Mercy of God which is over all His Workes, maketh the Grasse to grow upon Mountaines, and Herbes for the use of Men, and hath not onlie stamped upon them a distinct Forme, but also given them particular *Signatures*, whereby a Man may read, even in legible Characters, the Use of them." Thus, at its crudest, the theory was that plants with red flowers were good for the blood and yellow ones for jaundice; Scabious should be used for treating eczema because of its scaly buds and the mottled leaves of Pulmonaria were a sure sign that it was good for lung ailments. The theory was refined in the generation after Paracelsus by Giambattista Della Porta who taught that, through divine benevolence, medicinal plants had a similar distribution to the ailments which they would cure or relieve. In support of this notion it has often been pointed out that willows which are a source of salicylic acid, the basis of aspirin, do indeed grow in damp places where rheumatics and other agues are prevalent, but really the whole theory is a load of nonsense. The reasons it held credence for so long are not far to seek: most ailments clear up in time, whether they are treated or not; faith is also a great healer - faith, that is, in a worthless nostrum or placebo; doubtless many herbs which had long been known empirically to have curative properties were subsequently studied to discover their signatures; and finally, of course, people wanted to believe in a benevolent creator despite all the surrounding evidence to the contrary.

Anyway, progress did occur and the eighteenth century was less superstitious and more enlightened than the seventeenth. The great Swede Carl Linnaeus (1707 - 1778), who was at first a student of theology, early in his career held the orthodox view that God had created the species. When he was only twenty eight, he classified 7300 of them into 1098 families or genera. He adopted a binomial or two-name system which is essentially the one we use today, as John Ray had done before him in England. Later in life he concluded that God had made the genera, each originally with a single species, and that all the other species were the result of natural hybridization. He knew very well what spectacular results Dutch plant-breeders were obtaining by crossing plants. It is clear that he believed in a cautiously expressed kind of evolution that was contrary to Church teaching a century before Darwin.

His system of classifying plants was an ingenious but artificial one based on the number, arrangement and structure of their sexual organs - stamen, pistil, style and so on. He devised it initially for the plants of his own parish and regarded it as no more than a convenient pigeon-holing system for bringing some order into his branch of science. It was certainly an advance on grouping plants according to their supposed signatures, their size, or other earlier systems. It was adopted enthusiastically all over the world, making Linnaeus famous, nowhere more so than in England and by Charles Darwin's grandfather Erasmus. He celebrated it in his poem *The Botanic Garden* with thousands of coy rhymed couplets about the love-life of flowers.

The Linnaean system continued in use well into the nineteenth century though, with thousands of new plants being discovered every year, it was increasingly found to lump together strange and clearly ill-matched bedfellows, the Crocus and the tropical Tamarind tree, for example. A natural rather than an artificial grouping of plants was required, and major advances in this direction were made by a Frenchman, A.L. de Jussieu, and a Swiss father and son, Augustin and Alphonse de Candolle. The latter lived from 1806 to 1893. Once the evolution of species was accepted, attention was focussed on the underlying similarities of plant structure which closely related species show.

Alphonse de Candolle's common-sensical rules for standardising plant names were adopted by an International Botanical Congress in 1867, the first of its kind. They laid down that one plant species should have only one binomial name which should not be used for any other species. Where a plant already had two or more names, the earliest published - since 1753 - should be adopted (followed by its author's name): 1753 was the date of Linnaeus's *Species Plantarum* which therefore formed the basis and provided the pattern for all subsequent naming.

In Linnaeus's binomial system, the name of the plant's genus comes first and is given a capital letter. It is treated as if it was a singular Latin noun in the nominative or subject case no matter where it comes from. The specific epithet, which is usually an adjective, comes second without a capital letter and it has to agree with the noun according to the rules of Latin grammar. In Latin nouns have arbitrary gender - a bean is masculine, a carrot feminine and a turnip neuter - a tiresome irrelevance which English has sensibly done away with. Aster is masculine so the alpine Aster's specific name is *alpinus*; Clematis is feminine so the alpine Clematis is *alpina*; while Chrysanthemum is neuter and the alpine species is *alpinum*. Other common adjectives like *albus* (white), *hirsutus* (hairy), *longus* (long), *nanus* (dwarf), *parvus* (small), *robustus* (strong), *squalidus* (untidy, dingy), and *speciosus* (showy) all behave in the same way. Gardeners who learnt some Latin at school will remember that other nouns and adjectives behave somewhat differently, but this is not the place for a classics lesson.

The choice of Latin for naming plants was inevitable since it was the international language of scholars for many centuries and this was still true in Linnaeus's day. Unfortunately, a knowledge of school Latin and an ordinary Latin dictionary, though useful, do not go all the way in unravelling the meanings of botanical names. Some specific names are straightforward Latin adjectives like those above, but others are from Greek. Latin had to be adapted to describe the parts of plants; stamen literally means a thread or string and pistil a pestle. Other terms for a plant's anatomy are similarly adapted from Greek; calyx is a cup, style a pillar. Sometimes a specific name is a proper noun in the possessive case; Penstemons *barettiae* and *hartwegii* are named after the Mrs Barret and the Mr Hartweg who discovered them. Linnaeus, in his day, regarded it as a sacred duty to commemorate the names of botanists who had deserved well of the science. Generic names likewise come from several sources besides the classical languages: from other languages like Arabic or Chinese; from the names of botanists like Clarkia and Forsythia; from the names of places like Cydonia and Colchicum; from the names of real historical characters like Euphorbia and Gentian or from mythical ones like Artemesia and Narcissus. A good dictionary of plant names, which need not cost much, is essential for any gardener who is etymologically alert or has his curiosity whetted by this book.

Substantial agreement about all aspects of the rules for naming plants, now about 300,000 of them, was not reached until another Botanical Congress as recently as 1959 in Montreal. Revision of the names of plant families is even more recent and in some cases the old and new names are confusingly both being used. Final agreement about the *classification* of plants is another matter and still a long way off. That is why it often appears to the exasperated gardener that the experts still cannot agree about names and keep changing them for fun. Anyone over the age of fifty is likely to have grown up calling the Flowering Quince 'Cydonia' rather than 'Chaenomeles', talking about 'Montbretia, rather than 'Crocosmia', and getting his Pelargoniums and Geraniums in a twist. There are several reasons why there still appears to be confusion over plant names. New species are still being discovered which have to be named, of course, but that is not the main problem The main difficulty really goes back 100 million years to the Cretaceous period, long before today's continents and mountain ranges were formed, when hundreds of flowering plants appeared, in geological terms rather suddenly. Some resembled plants we know today, like the Magnolias and Water Lilies; others became extinct long ago. The fossil record is full of gaps and puzzles; the genealogy of plants - who is descended from whom - is far from fully understood. New evidence comes to light and leads to revisions. Evolution, too, is not something that happened a long time ago: it is still going on and the plant world is now, as it always has been, in a state of flux, much accelerated today by man's activities. Some species, especially some that occur over a wide geographical area, turn out to be highly variable, like a collection of subspecies with no one obvious type of which they are variations. We are also now able to look inside a plant's cells and study its relationships at a biochemical level. The result of all this is that some species get moved from one genus to another when evidence emerges that that is where they really belong. Less often, a whole genus gets shunted from one family into another. Some groups of plants which were once thought to belong together have had to be split up; others which were once thought distinct are now lumped together. The process will go on and all the gardener can do is learn to adapt to the consequent changes of name. There is a good reason for them.

Many familiar garden plants do not occur anywhere in the wild and have been in cultivation for so long that their ancestry is obscure. In addition, as gardeners know very well, there are thousands of modern 'cultivars', hybrids, varieties, forms, strains and so on deliberately created by plant-breeders and this adds a whole new dimension to the problem of plant-naming. There is an International Code of Nomenclature for Cultivated Plants (the Cultivated Code, for short) which was agreed in 1952 and has to be revised frequently. A cultivar must retain its characteristics when propagated. The first person to describe a new one in a reputable publication has the right to name it according to the rules of the code. That name will follow the botanical name, be in 'common language' rather than Latin, have a capital letter and either be in inverted commas or printed in a different type, often preceded by 'cv'. So, if you want the splendid golden form of the common False Acacia or Black Locust Tree, you will look out for *Robinia pseudoacacia* cv. 'Frisia'. The Kalmia is known as the Calico Bush or Mountain Laurel but if you seek the best variety for your garden, you will order *Kalmia latifolia* cv. 'Clementine Churchill'.

Some plant retailers, particularly in the mail-order business, use eye-catching fancy names which have more to do with the fertile brains of advertising men than with the Cultivated Code. They are not all crooks but, if you order from them, you will not know precisely what you are getting: it is likely to be something cheap and ordinary which will not transform your garden like magic or make your neighbours gasp with envious astonishment. It probably ought to be labelled 'Caveat emptor' or Buyer-beware', two useful binomials! If we are honest, most of us have been once bitten and that is how such businesses survive.

I chose the example of Robinia and Kalmia partly because they both have at least two common English names. That is the problem with 'common', vernacular or English names, delightful and descriptive though they often are: one plant may have many different names and, even more confusingly, one name may be used for many different and quite unrelated plants. We had better take a deep breath and then start a new chapter about them.

CHAPTER 3

PLANTS GET COMMON NAMES

I have on my shelves a Victorian tome by R.C.A. Prior M.D. with the catchy title:- *On the POPULAR NAMES of BRITISH PLANTS being an explanation of the ORIGIN AND MEANING of the names of our INDIGENOUS AND MOST COMMONLY CULTIVATED SPECIES.* As one would expect, the good Doctor is evasive about some of the coarser country names and many of his explanations would not be accepted by scholars today: nevertheless, his book is full of interesting information. He quotes quite often from the Greek Herbal of Theophrastus, from Pliny's Natural History and from the Tudor herbalists. He starts with Aaron, a misspelling of Arum as a name for Lords-and-Ladies, and ends up some 250 pages later with Youthwort, a mistake for Ewe's-rot, an old name for Round-leaved Sundew which was once thought to cause foot-rot in sheep. Both these examples will remind us of the process known as 'folk etymology' which we discussed in the first chapter, and that the common names of plants are not always what they seem at first sight.

The main problem with them, as we were saying at the end of the last chapter, is their multiplicity and, hence, their ambiguity. The scientific name *Caltha palustris* refers to one species only and is used for no other. It means 'Marigold of the Marsh' and I suppose Marsh Marigold is the nearest we have to a standard English name, though it is a member of the Buttercup family and not remotely related to the French or African Marigolds (*Tagetes*) or to the Pot Marigolds (*Calendula*) which are all Composite flowers. Actually I grew up calling them Kingcups, but in some places that name is used for other Buttercups, and there are many other names for Marsh Marigolds like Bull-cups, Butter Blobs, Gowans, Marsh Lilies and so on. Indeed there are said to be eighty names for them in English, nearly as many in France and twice the number in Germany. When a great deal more of the countryside was ill-drained and marshy than is the case today, the Marsh Marigolds in flower must have been one of the commonest and most spectacular sights of Spring and so they have attracted many names. All kinds of things have regional or dialect names from the weakest piglet in the litter to left-handedness, and conspicuous wild flowers are no exception.

If we take another of our favourite Spring flowers, the question of names is even more complicated. What we call a Bluebell in England (though it has many other country names like Greggles or Snap-grass) means a Harebell (*Campanula rotundifolia*) in Scotland, where they call our Bluebell 'Wild Hyacinth' (and other things). Now the Hyacinth is named after a character in Greek mythology, a young prince of Sparta. He was accidentally killed by the discus of Apollo who, grieving, caused flowers to spring from his drops of blood. Gerard and his contemporaries were unsure whether our Bluebell was the Hyacinth of classical writers: they call it the English Jacinth or Crowtoes and say its bulb was used for making starch and glue. There was uncertainty, too, among those classical writers whether the plant whose leaves were supposed to be inscribed 'Alas! Alas!' really commemorated Hyacinthos or the hero Ajax of Salamis. In any case, it was not our Bluebell the Greeks knew, which only grows along the Atlantic edge of Europe. Some think it was the wild Gladiolus, others that it was one of the Mediterranean Scillas. A plant which springs from the blood of a hero ought to be what Milton calls 'a sanguine flower inscribed with woe'. Linnaeus could see that our Bluebell was not blood-coloured so he named it after another mythological Greek. He called it *Endymion* after the beautiful youth who was kissed in his sleep by Selene, the moon, and added the specific name *non-scriptus*, 'not inscribed (with marks of woe)'. Today we have kept his specific name but changed the genus to *Hyacinthoides*.

There remains the adjective 'hyacinthine': Geoffrey Grigson says, "The proper hyacinthine colour is blue - the colour of the precious stone *huakinthos* (probably the aquamarine)." Other authors disagree, saying that the precious stone so named - including the Jacinth beloved of poets - was a clear brownish-red form of Zircon. Take your pick. One further trap remains to be sprung. If, in poetic mood, you describe someone's hair as hyacinthine, it means neither bloody locks nor yet a blue rinse: it is remembering Homer's description of Hyacinth's tresses which so attracted Apollo as 'like silver overlaid with gold by a skilled craftsman', what we would call ash-blond perhaps.

If you remember all that, it will enable you to be very erudite about Bluebells. It may not enhance, but neither will it detract from your pleasure in the sight and scent of them in flower. These two, Marsh Marigold and Bluebell, are not exceptional. Think of Celandines: the Lesser is a Buttercup, sometimes called Pilewort for reasons to be explained elsewhere; the Greater is a Poppy. Only the latter should really be called Celandine which comes from the Greek name for swallow. Pliny says swallows use the juice of the plant to restore the sight of their nestlings even when their eyes have been picked out, and Poppies all have a milky sap or Pap (hence the Latin name Papaver from which Poppy derives). How a sensible man gave credence to such an extraordinary idea and where it can have come from are hard to imagine.

Another improbable name which comes from one of the more fanciful ideas of Pliny's day is Loosetrife. We have as native plants Yellow Loosesrife (*Lysimachia vulgaris*), the tall relative of Creeping Jenny, and the unrelated Purple Loosesrife (*Lythrum salicaria*). The name was given to both by the Tudor herbalist William Turner who was translating Pliny's *Lysimachia*, which means 'deliverance from conflict'. Pliny appears to give two reasons for the name: it was discovered, he says, by Lysimachus, a companion of King Alexander of Thrace who lived about 300 B.C. but he also says that, placed on the yoke of fractious oxen, it will stop their strife.

And so we could go on. Fascinating? That depends on your point of view. English names in their multiplicity and diverse origins are certainly a muddle. One can understand the botanist, the horticulturalist and even the tidy-minded amateur gardener who want as little to do with them as possible. We could do worse than refer them to Doctor Prior's book with which we began, for some words which are as sensible today as when he wrote them in 1870. He says, "There are distinguished botanists at the present day who look upon popular names as leading to confusion, and a nuisance, and who would gladly abandon them, and ignore their existence. But this is surely a mistake, for there will always be ladies and others, who, with the greatest zeal for the pursuit of Natural History, have not had the opportunity of learning Greek and Latin, or have forgotten it, and will prefer to call a plant by a name that they can pronounce and recollect ... Besides, admitting to the full all that can be urged against them from a purely botanical point of view, we may still derive both pleasure and instruction from tracing them back to their origin, and reading in them the habits and opinions of a former age."

When we take the good doctor's advice and trace plant names back to their origin, we will sometimes find that they have come down to us as part of the language itself. They go back to pre-literate, pre-historic times, to the Indo-European language from which most of our European and some Middle Eastern and Asian languages have evolved. This is true of the names of many of our native trees like Ash, Beech, Holly, Oak and Willow. They all have 'cognate' or related names in other branches of this Indo-European language tree, particularly in the Germanic languages of northern Europe. Our Apple, for example, is *Apfel* in German, *afal* in Welsh, *ubhall* in Irish and *aval* in Breton: it crops up in the name of the town Avallon in France, in King Arthur's Avalon, wherever that was, as well as in English place-names like Appleby and Appledore. It presumably referred to a wild or crab-apple; you cannot go any further than that though you can, of course, explain the names of cultivated apples like Bramley or Russet. Our Lime or Linden is *lind* in Swedish and that is where the great Linnaeus's name comes from.

Many of our plant names are not as ancient as Ash or Apple; they have been given more recently and more deliberately and these are the ones which will tell us about 'the habits and opinions of a former age'. The name Dyer's Greenweed (*Genista tinctoria*) tells us that it was formerly used for dying cloth: it is one of our native Brooms and other members of the genus were obviously used for making brooms for sweeping. Feverfew (*Chrysanthemum parthenium*) from medieval Latin *febrifugia* was clearly believed to put fever to flight. These are both names from the real or imagined use to which the plants could be put. Other names came from the appearance of plants, from their behaviour, habitat, place of origin, taste or smell or from old legends or myths like that of Apollo and Hyacinth.

Let us now give a few examples of each of these name-sources, choosing specimens mainly from our native flora: garden plants, which are often introductions from elsewhere in the world, will get the lion's share of attention in later chapters, and wild gardens are now much in vogue in these environmentally concerned days.

Of Physic, Fleas And Nostrums

When life was nasty, brutish and short for many people, it is unsurprising that they cared little for wild plants if they were found to be neither harmful nor useful in improving their lot in one way or another. Apart from the trees of the forest which were essential for fuel, timber for building, pannage for pigs and a hundred other purposes, they tended to divide the vegetable world into pot-herbs for food, nose-herbs to strew about for their scent, and medicinal herbs for curing their ailments. They ate wild fruits in due season and herbs to pep up a dull dish, but they were not, I suspect, great vegetable eaters except in dire necessity; bread was the staff of life with a bit of meat when you could get it. Salads were eaten mainly for their laxative properties. Very few of our native plants have names derived from their use as pot-herbs. Corn Salad (*Valerianella locusta*), also known as Lamb's Lettuce, and Salad Burnet (*Poterium sanguisorba*) were presumably eaten raw then as now, and 'cress' comes from an ancient root-word meaning to eat or nibble.

Of strewing herbs, the Bedstraws' name indicates their use (unless it was originally 'bead-straw') though *Galium*, their generic name, means 'milk' - which they were once used to curdle in cheese-making. Lady's Bedstraw (*G. verum*), Woodruff (*Asperula odorata*) - Asperula means 'little rough one' - and Sweet Vernal Grass (*Anthoxanthum odoratum*) are all more strongly scented when dried than fresh. Medieval houses with their lack of sanitation must have badly needed their sweetening fragrance. Sweet Flag (*Acorus calamus*) is an introduction but, by the seventeenth century, it was common enough in the Fens of eastern England to be used to carpet

Norwich Cathedral. Fleabane (*Pulicaria dysenterica*) must have been useful too. It is called "an ill-looking weed" by Culpeper, who says that its smell is delightful to insects but its juice destructive of them, which makes me uncertain how it was used. I thought I had read of it being scattered among bedding and also hung in bunches at the window to keep out midges. Like Bedstraw, its Latin name tells that it also had two uses; *Pulicaria* is from *pulex* - a flea, but it was also used for treating dysentery.

The number of herbs once used medicinally is very large; there are about 400 in Culpeper's herbal; and quite a few have names to match their use. How efficacious they were is debatable since science and superstition are frequently inseparable. Scurvy was not a scourge of sailors only: even royal persons who did not eat their greens could be smitten with "this filthie, lothsome, heavie and dull disease" as Gerard calls it. Henry VIII is now thought by some to have died of it. It seems that Dutch mariners were the first to use Scurvy-Grass (*Cochlearia officinalis*), which grows conveniently on the sea shore, to prevent it. Others soon caught on to its antiscorbutic properties which are real and not imaginary. It continued in use until lime-juice was found to be a pleasanter tasting and equally convenient source of Vitamin C to take on a voyage. The proper name *Cochlearia* is from the Greek for a snail-shell, because of the whorled shape of the plant's seeds.

Considerably more dubious is the case of Wormwood (*Artemisia absinthium*) which is *Wermut* in German and *Vermouth* in French. Its bitter taste was known to the Greeks, who called it *Apsinthion* and it is still used to flavour that potent green liqueur. 'Artemisia' was the name of a Queen of Caria in Asia Minor and she was named after Artemis, the Greek name for Diana, daughter of Jove, sister of Apollo and Goddess of hunting and the moon. In Old English Wormwood is *wermod*; its etymology is unclear though it probably has nothing to do with either 'worm' or 'wood'. Bitter herbs were believed to help rid the body of worms and that is how the name developed - another instance of folk-etymology. Culpeper says it also helps hysterics and obstructions of the spleen, while old Thomas Tusser in his *Five Hundred Points of Good Husbandry* of

1573 recommends scattering Wormwood seeds rather than Fleabane in the bedchamber "to make flea to refrain". Of July he writes:-

"While Wormwood hath seed, get a bundle or twain,
to save against March, to make flea to refrain:
Where chamber is swept, and Wormwood is strown
no flea, for his life, dare abide to be known."

The value of herbs which were supposed to heal wounds like Self-Heal (*Prunella vulgaris*) and Woundwort (*Stachys palustris*) is similarly doubtful though Gerard has several stories from his own experience of them curing dreadful injuries. In his picturesque English he says that, when made into a poultice, they "stay the flux of blood from wounds and solder up their lips". However, a hook was thought to be the signature of a vulnerary herb and, in both, the upper lip of the calyx is hook-shaped. In neither case does the Latin name suggest a surgical application: *Prunella* suggests 'little plum-coloured one', not a bad name for Self-Heal, and *Stachys* is a spike: probably a poultice of grass would have done as well, or of Salad Burnet, mentioned earlier, whose specific epithet is *sanguisorba* or 'blood-stauncher'.

From the dafter, more superstitious end of the spectrum of use-names, we will just take three more examples. Our handsome Viper's Bugloss is *Echium vulgare* because it was identified with the Greek *ekhion* (meaning a viper) of Dioscorides, which had seeds he thought resembled a serpent's head. The plant was consequently supposed to cure snake-bite. Bugloss is a better name: it also comes from Greek, meaning 'ox-tongued', which is descriptive of the plant's rough, bristly leaves.

Nipplewort (*Lapsana communis*) is common everywhere and the name that Parkinson coined for it seems to be the common one. There are a few other country names like Hasty Rogers and Carpenter's Apron from the shape of the leaves, but it is the shape of the buds which led the herbalists to call it *Papillaris* or *Herbe aux mamelles*. Actually other members of the same family like Sow Thistle (*Sonchus asper*) seem to my eye to have more distinctly nipple-shaped buds, but *Lapsana* was evidently the one that was made into a salve or unction for the breasts.

Lastly, let us take Figwort (*Scrophularia nodosa*), a handsome if somewhat smelly plant. *Nodosa*, which can mean 'jointed' or in this case 'knotty', is the subterranean clue to its names in English and in Latin and to its imagined use. It has little nodules on its rhizomes, which were read as a sign that it would be useful for treating King's Evil or Scrofula - tubercular swellings of the lymphatic glands in the neck. (Scrofula comes ultimately from Latin *scrofa*, a breeding sow, but don't ask me why.) It was also thought to cure piles: Latin *ficus* means both fig and also haemorrhoids, which were off-puttingly thought to resemble that fruit.

Many of these names, in themselves, are good descriptive ones derived from close observation: it is the deductions about the plants' medicinal uses that are often sad and silly in their pre-scientific ignorance and their reminder of the many ills that afflicted our ancestors. Observations of the attractions and repulsions of certain plants to domestic birds and animals have more sense in them. Cats certainly like *Nepeta cataria*, pigs relish *Heracleum sphondylium* which is known as Hogweed or Cow-Parsnip, and Henbane (*Hyoscyamus niger*), which seems particularly at home round the untidy edges of farm yards, is poisonous to poultry.

Of Clovers, Clubs and Coltsfeet

What is the connection between Hercules, a pack of cards and St Patrick? Even if you were thinking along the lines of plant names derived from their appearance, you might take some time to solve this riddle. Hercules is usually depicted with his *claver trinoides*, the huge three-lobed club which he cut for himself in Nemea and with which he performed many of his heroic feats of strength. The claver or club gives us both the plant name Clover and the fourth suit in a pack of cards, but where does St Patrick fit in?

The only time I have had my face slapped was when I innocently asked a waitress why she had a piece of Watercress pinned to her apron. I was somewhat aggrieved by her volatility because I did not know that it was St Patrick's day or that she was Irish. I still think it was Watercress too, and I have Tudor writers on my side. The chronicler Holinshed, talking of the Irish, writes of "watercresses which they tearme shamrocks" and Fynes Morison, who evidently had a low opinion of them, says "they willingly eate the herbe shamrock, being of a sharp taste, which as they run to and fro, they snatch like beasts out of ditches". How this mistake arose I do not know, but mistake it must be. Irish *seamrog* means 'little clover', Watercress is not a trefoil and yet it must have been a trefoil which St Patrick used to illustrate the doctrine of the Trinity. But which trefoil is the true Shamrock? Blood has probably been spilt over that question. An Irish botanist, Nathaniel Colgan, seeking to solve it once and for all in 1892 invited examples to be sent to him from all over Ireland. What he received were specimens of Spotted Medic, Small Medic, Hop Trefoil and Red and White Clover, but *not* Wood Sorrel, which has sometimes been urged as a candidate. The answer is evidently that different trefoils are called Shamrock in different parts of Ireland, though there is a marginal majority in favour of Spotted Medic or Nonsuch (*Medicago arabica* which has also been called *maculata* in the past).

It seemed worth clearing up that vexed question as far as we can, even if it has been a long-winded way of making the point that many plant names come from their appearance rather than their use, often from leaf colour or shape as in the case of clover. Coltsfoot, Dandelion (French *dent de lion*), Silverweed, Cinquefoil and Twayblade are other obvious examples. Sometimes the whole plant suggests the name as, I take it, in the case of Jacob's Ladder, Spiderwort (*Tradescantia*) whose sprawling jointed stems suggest spiders' legs and *Petunia*, named because of its resemblance to the Tobacco plant which is *petun* in French.

Often a resemblance is seen between the flower or some part of it and something else. Sometimes the resemblance is striking enough, as with the Bee Orchid, to fool a pollinating insect into thinking it is copulating with one of its own species. Townhall Clock (*Adoxa moschatellina*) with its four faces and Star of Bethlehem (*Ornithogalum umbellatum*) are apt and charming but the tiny

Milkwort (*Polygala vulgaris*) less obviously has an udder-shaped flower which must have suggested that cows who ate it would produce *poly gala* (much milk). Columbine is from Latin *columba*, pigeon, because the petals resemble five doves, but the scientific name Aquilegia shows that they can also be seen as resembling eagles (*aquillae*). Fritillary (*Fritillaria meleagris*), which is sometimes well named 'Snake's Head' in English, is from Latin *fritillus*, a dice-box which opens out into a chequerboard, while *meleagris* refers to the similarly chequered plumage of a Guinea-fowl.

The seed-pod or carpel which follows the flower can be an equally distinctive feature and gives names like Shepherd's Purse, Bird's-foot Trefoil and Honesty (*Lunaria annua*) because you can see how many coins it has in its transparent, moon-like (lunar) purse. And there are stem names like Knotgrass and root names like Coral-Root, Bird's-Nest Orchid and Devil's-Bit Scabious, whose abruptly ending rootstock looks as if it had been bitten off. Those old herbalists looking for signatures and cures were certainly close observers.

Of Cowslips, Cleavers And Cuckoo-Flowers

There is another group of flower names which could be said to derive from their behaviour, that is to say how, where and when they grow. Let us take 'how' names first. There are those interesting plants with no green parts of their own, the Broomrapes, which manage to grow by parasitising other plants. Not only Brooms are attacked but also Gorse, Thyme, Bedstraw, Yarrow, Hawk's-Beard and others. Their Latin name *Orobanche* means 'legume-strangler' but, in this country, there is no species which attacks cultivated members of the pea family.

It is surprising that our two native insectivorous plants, the Butterworts and Sundews, do not have many common names which refer to their carnivorous habits. Butterwort was apparently once used to curdle milk and also had magical properties in protecting the udders of cows but Bog Violet is, perhaps, a better name for it since its insect-catching activities are not too obvious. Sundew plainly catches midges and other small creatures on its sticky paws and is known as Fly-Trap in Devonshire.

Sunflowers and Heliotropes turn with the sun (Greek *helios*), Daisies (day's eyes) open and shut with the daylight and Pansies (French *pensée*) hold their heads at a thoughtful angle. Periwinkle is from Latin *per vincula*, to bind about, and the name Vetch comes from the same verb. Vetches were the Tares of the Bible and a cornfield curse in the days of hand-reaping. Cleavers or Goose-Grass (*Galium aparine*) cleaves and climbs by means of downward-curving thorns so tiny that the plant feels sticky rather than prickly: did it inspire the inventor of Velcro? It could well have done.

When we come to 'where' names, we have both habitat and place-of-origin names. A Wallflower is particularly happy growing on an old wall, but a Walnut needs more explaining. The Welsh, it must be remembered, call themselves Cymry: it was the Anglo Saxons who called the Britons they found in these islands the Welsh, which is German *wälsch* and meant 'strange, foreign, outlandish'. They had no high regard for the dark and dwarfish people they had conquered and enslaved; and the places where they lived, if they had not fled to the mountains, often have names like Walton, Walmer, Walworth and Wallasey. So a Walnut is a strange, foreign nut just as people who got the names Walsh or Wallace were strangers from the west or north. The Welsh Poppy (*Meconopsis cambrica*) really does come from Wales, but the Welsh Onion has no more to do with the Principality than the Walnut: it's just outlandish.

Many of our cultivated flowers, fruit and vegetables are named after the places they came from, Sprouts from Brussels and so on, and a few of our rarer natives have names from the places where they grow like the Portland Spurge and the even rarer Cheddar Pink which only grows on the cliffs of the gorge where it was first discovered in 1723. Canterbury Bells seem to have been named by Gerard who says they grow very plentifully in low woods about that city. Another theory is that they resemble little bells which were sold as souvenirs to pilgrims to the shrine of Thomas à Becket. *Campanula*, of course, means 'little bell'. I doubt whether Gerard was talking about our Canterbury Bells (*Campanula medium*) which come from Southern Europe. More likely he had seen a native like

Nettle-Leaved Bellflower which is handsome enough to grace any garden. So would be Grass of Parnassus (*Parnassia palustris*) if you could get it to grow there: the name may not be wholly fanciful since it grows in marshy places through much of Europe and Asia, but as far south as Greece it becomes a mountain plant.

No one would call Oxford Ragwort (*Senecio squalidus*) a handsome plant. It is a coarse relative of Groundsel and is no more a native here than Canadian Fleabane (*Erigeron canadensis*)though both are now widespread weeds. *S. squalidus* should really be called Volcano Weed since it is at home on the cinders of Vesuvius and Etna. Its story has been told many times but will bear repeating because it illustrates the importance of the right habitat to a plant's flourishing. A specimen was growing in the Oxford Botanical Garden at the end of the seventeenth century and Linnaeus saw it there when he visited in the next century. In 1794 it escaped and was seen growing (as it still does) on the walls of some of the college buildings. Eventually in 1877 it reached the railway and, nearly two centuries after it was introduced from its volcanic home, it felt happy on the cindery ballast of the tracks and started to spread. By 1890 it had reached Swindon and then there was no stopping its march along the railways all over the British Isles. During the Second World War it found bombed sites almost as congenial as the slopes of Vesuvius. I saw it growing on Platform Six of Crewe's modern station just the other day.

Habitat is very obviously suggested in binomial names like Sea Pink, Bog Asphodel, Marsh Gentian, Field Scabious, Wood Sorrel, Alpine Lady's Mantle, Rockrose and so on. Moss grows in mosses or marshes and so does Cranberry which is where cranes would also have been found. They were once common birds in Britain as one can tell by the frequency of place-names like Cranbrook, Cranford, Cranmere, Cranwell and several others.

Cowslip and the sturdier Oxlip must also count as having habitat names. If you thought they had something to do with cow's lips, you were probably misled by one of those polite Victorian floras. Even Doctor Prior could not accept that Old English *cuslyppa* really meant what it said and goes off into a rigmarole about the name originally applying to Mullein and meaning 'hose-flap'. A cowslip is a cow-slop or cow-pat, what the Americans rather coyly call a 'pasture platter', and it was once believed that these plants only grew where cattle had been. "There is", as Dr Prior says, "very little poetry in some of these popular names."

Lastly in this section, mention must be made of 'when' names, which nearly always refer to the time of a plant's flowering when it is most conspicuous. Spring and Autumn Squills, lovely on Cornish cliffs, Autumn Crocus and Winter Aconite are obvious candidates. Several plants have been called Cuckoo-Flower and Swallow-Wort from flowering when those birds arrive. Lent-Lilies are Daffodils, in flower well before Shrove Tuesday (as I write in an exceptional year), and Pasque Flowers (Pulsatilla) are Easter Anemones from the Old French *pasques* for Easter.

Fluellen baffled me for some time because it is the name I know for the little creeping Toadflax of cornfields (*Kickxia elatine*), yet I had read that it was named after St Llewellyn whose day is April 7th and much too early. Dr Prior says the name comes from the Dutch word meaning 'downy' or 'velvety', not from the Welsh saint, but he distinguished between Male Fluellen, the creeping Toadflax which does not appear at all downy to me, and Female Fluellen, the Common or Germander Speedwells which certainly do have hairy leaves and might be found in flower as early as April. There is a puzzle here and, surprisingly, Geoffrey Grigson who can usually be relied on to sort out such matters does not include the name in his *Englishman's Flora*.

After April comes May-flower on the Hawthorn and the very rare May Lily (*Maianthemum bifolium*) in a few woods in Lincolnshire and Yorkshire, and then the St John's Worts which are in flower around his feast day, which is June 24th.

Of Mustard and Musk

Only a handful of plants get their common names from their taste or smell and then usually as a qualifying adjective as in Sweet Cicely, Bitter Cress and Fetid Iris (Gaddon). The Stinking Hellebore has an astringent breath which is far from unpleasant to my nose and, like its pale green panicle of purple-tipped flowers, a quintessential part of early spring in chalk country. It is not a smell that tempts one to try eating the plant, however, which is just as well: Hellebore means 'poison-food'.

Jack-by-the-Hedge (*Alliaria petiolata*), though edible, is more of a stinker particularly when crushed. Dr Prior, last century, said the name came from jakes or privy and Dr Denis Owen repeats this idea in his recent book *What's In A Name* based on his broadcast talks. The first was a medical doctor and the second is an ecologist: to a mere etymologist, however, the suggestion seems unnecessarily ingenious. There are many Jack- names (see under Jackdaw in the chapter on garden birds) with no implication of an offensive smell. *Alliaria* means 'garlic-like' and other names like Garlic Mustard and Sauce-Alone show that it had its culinary uses. It was made into a spring sauce and often served with salt fish. Culpeper, who says it smells more pleasantly than garlic, recommends it for warming the stomach and aiding digestion.

Another green-sauce plant is Common Sorrel (*Rumex acetosa*) which was once widely used in cooking for its acidity before lemons were available. 'Sorrel' is from Old French for 'little sour one' and most of the English country names are similar: Sour Docken or Sourlick for example. Culpeper says it cools the blood, quenches thirst and produces an appetite in 'fainting and decaying stomachs'.

For sweeter things we might turn to Roseroot, Musk and Chamomile. Roseroot (*Sedum rosea*) has a name which is precise in Latin and English, though the flowers are yellow and not pink as you might expect. Sedums sit (from Latin *sedeo*) like green cushions on rocks and this particular frequenter of mountain ledges has a tough, thick stem which, improbably, has a delicious rose-like scent when it is cut or bruised.

Musk (*Mimulus moschatus*) was introduced to gardens here from north-west America early last century, soon established itself along our waterways and then mysteriously lost its scent. Latin *moschus* is musk and *Mimulus* is 'ape - flower' or 'monkey - flower'. There is a connection between to ape, to mimic and to mime and Mimosa gets its name by folding its leaves at a touch like the silent gesture of a mime-actor or *mimus*.

The wild Chamomile (*Matricaria recutita*) has not lost its scent, but then it has been here as a cornfield weed for a very long time and seems as good to me for fragrance (but not for making a lawn) as the cultivated *Chamaemelum nobile*. All manner of benefits are claimed for drinking Chamomile tea from chastening disobedient rabbits to raising downcast spirits and it has been used for shampoos and poultices. *Matricaria* indicates that it treated ailments of motherhood and *Chamaemelum* is from Greek for 'ground-apple' from its carpeting habit and apple-scent. To our noses it smells even more like Pineapples, which were not known in the ancient world since they are native to Central America. I have come across places called Pineapple Farm or Pineapple Field when walking in several counties or when map-browsing, and they got their names, I feel sure, from one of the Chamomiles. Again and again I have read that gorse flowers smell of Almond whereas anyone who has sniffed with unprejudiced nose will agree that the scent is much closer to Coconut. Could this be another case of the smell being described when Almonds were known but Coconuts were not? The Portuguese first brought Coconuts to Europe in the sixteenth century and called them *coquos* meaning 'grinning faces' and that is presumably why we have Coco the Clown. One thing is for sure: it is tempting for writers of books to copy from each other without going out and checking their facts with their own senses. It has happened since the earliest times and the most ardent outdoor man would have to plead guilty to being a closet-naturalist occasionally.

Leonhart Fuchs 1501 - 66
He produced the most beautifully illustrated of all the early herbals and had the fuchsias named in his memory.

Chapter 4
Buddle, Fuchs & Co.

All gardeners know moments of horticultural exasperation when, as W.S. Gilbert puts it, "I conceive you may use any language you choose to indulge in without impropriety". When strong language seems called for, my favourite oath is "Oh Buddle and Fuchs!" It is powerfully cathartic - try it - and yet even a bishop could utter it without a blush. Indeed Adam Buddle was a reverend gentleman, though not a bishop, and I have no grudge against him apart from the fact that he had nothing whatever to do with Buddleia (in fact he was an expert on mosses). And Leonhard Fuchs produced the most beautifully illustrated of all the early herbals in 1542, 150 years before he could have seen the first Fuchsia (we should really pronounce it 'Fooksia') which was introduced from Mexico at the end of the seventeenth century. How come? You may well ask.

Scientific names from Greek and Latin are bad enough: they are undoubtedly necessary and usually have some descriptive sense in them when they are translated, but there is no getting away from the fact that they are not often easy on the eye or the ear of the English gardener. Edward Lear was too much of a traveller to be a great gardener, but he made fun of those names in his *Nonsense Botany* of 1870. Can you picture his *Nasticreechia krorluppia*, his *Phatfacia stupenda* or his *Barkia howlaloudia*? If not, you had a deprived childhood. He does not seem to have noticed the practice of naming plants after people like Buddle and Fuchs or he would surely have made fun of it. It did not escape the attention of Oscar Wilde, however. He gives Lord Henry in *The Picture of Dorian Gray* these remarks: "Yesterday I cut an orchid for my buttonhole ... In a thoughtless moment I asked one of the gardeners what it was called. He told me it was a fine specimen of *Robinsoniana*, or something dreadful of that kind. It is a sad truth but we have lost the faculty of giving lovely names to things."

I have much sympathy with Lord Henry: the proper place for the names of dead botanists is on their books, their tombstones and perhaps on the houses where they once lived. However, Linnaeus, the greatest namer of them all, approved the practice of giving them to plants, though he did not originate it. In his *Critica Botanica*, he wrote, "Generic names formed to preserve the memory of a botanist who has deserved well of the science I retain as a religious duty." And so we are stuck with them, scores of them. We have been surprisingly slow to invent good English names for most of them.

It was Charles Plumier who decided in 1702 to commemorate Fuchs, but he could hardly have chosen a less happy name for those dainty, pendulous flowers with their ballerina skirts. The hardiest variety of our West Country hedges is additionally called *magellanica*, thus also commemorating the great Portuguese navigator Ferdinand Magellan. Now Fuchs, as we were saying, was a worthy fellow. As well as his herbal he gave us 'Digitalis' as the name of the Foxglove genus. He became famous for saving many lives in an epidemic of the 'English Sweating Sickness' at Anspach, where he was physician to the Margrave, and he was later Professor of Medicine at Wittemburg. They should have put up a statue to him but, I suppose, it was cheaper to name a flower after him.

The Reverend Mr Buddle was a deserving character too. He died in 1715 and his laboriously collected herbarium eventually passed to the British Museum. It was Linnaeus himself who decided to honour Buddle's memory when the first Buddleia was brought from Chile or Peru in 1774. This was *B. globosa*, the Orange Ball Tree. The better known Butterfly Bush with its arching sprays of honey-scented purple flowers was collected in China much later by a Jesuit missionary, Père Jean André Soulié, who was tortured and murdered by Tibetan monks for his trouble. He did deserve a memorial. In China they call the shrub 'Summer Lilac' and thickets of it were said to be 'famous harbourage for leopards', which paints a magical miniature picture in the mind. It did not arrive in English gardens until the 1890s.

So quite a few of the genera which have personal names got them in the eighteenth century in memory of defunct botanists who never set eyes on them. Many of them were Scandinavians associated with Linnaeus, or Germans or Poles and unfortunately they did not always have euphonious names: *Rechstieneria*, I ask you!

Grabowskia is a spiny shrub and sounds it; but the Calla Lily is really out of luck in being lumbered with *Zantedeschia* (after an Italian, actually) and so is the Christmas Cactus with *Schlumbergera* (after a Belgian). Later on, as we shall see, some other genera were given the names of the explorers who first discovered them, just as rivers, mountains and, indeed, countries have been named. Many more specific names originated in this way. Sometimes it is the horticulturist or nurseryman who introduced or popularised the plant (like Jackman of the Clematis, Pissard of the Prunus or Lawson of the Cyprus) who is honoured: sometimes it was a patron of botany or other person of influence who was worth flattering.

A few plants were named after real or, more frequently, mythical persons in ancient times: Paeony, for instance, from Paion, physician to the gods on Mount Olympus, and the Pleione Orchids from the wife of Atlas and mother of the Pleiades. The Gentians were named by Pliny after a King Gentius of Illyria who was supposed to have discovered the medicinal properties of the tall alpine *G. lutea*: its root is still the source of the 'bitters' used to flavour some alcoholic concoctions. Names of this kind are not very numerous and are not included in this chapter.

Finally there are one or two oddities of which Sequoia, the Giant Redwood of California, is, perhaps, the oddest. 'Sequoiah' is a Cherokee Indian word which means 'opossum' and was also a slang term in that language for a half-caste. It was the nickname of George Gist (1770-1843) whose mother was a Cherokee and whose father was a British merchant of German extraction. He was brought up as a full-blooded Red Indian, later becoming a considerable scholar who devised the first syllabary for writing the Cherokee language so that they were one of the first tribes to be literate. The noble trees were named in his honour.

Here now is an A to Z of some (but by no means all) of the genera which got personal names in one way or another. I have tried to include those most likely to be known to British gardeners though not all provide hardy species, and I have reserved a clutch of climbers for the chapter of that name viz. Banksia, Bignonia, Bougainvillea, Garrya, Lonicera, Thunbergia and Wisteria. Others will be found in the chapter on bulbs and elsewhere.

AUBRIETIA, Purple Rock-Cress, was introduced from the Levant where it was discovered by J.P. de Tournefort and named after his artist-friend and travelling companion Claude Aubriet (1665-1742).

BEGONIA, Elephant's Ear, Beefsteak Geranium. The genus of about 900 species, all tender, is named after Michel Bégon (1638-1710), a patron of botany, who was Superintendent at St Domingo and later Governor General of French Canada.

BROWALLIA is generally known as a violet-flowered pot-plant though there are other species. It is named after Johan Browall (1707-1755), Bishop of Abo, a Swedish botanist who, unlike many churchmen of the time, approved of Linnaeus's sexual system of classifying plants.

BRUNFELSIA. Like Browallia, they are mainly greenhouse subjects and both belong to the Solanum family together with potatoes, tomatoes and Deadly Nightshade. Otto Brunfels (1489-1534) was a German monk who made more detailed and accurate drawings of plants than any of his predecessors.

BUDDLEIA after Adam Buddle (1660-1715). See above.

CAMELLIA. They are members of the tea family, it is true, but why will people pronounce the name 'Cameelia'? They are named after the Jesuit botanist Georg Joseph Kamel or Camellus (1661-1706) - pronounced 'camel' - who travelled in Asia, corresponded with John Ray in England and described the flora of Luzon in the Philippines.

CARPENTERIA. There is only one species of this evergreen shrub from California with scented, anemone-shaped white flowers. It is named after Professor William M. Carpenter (1811-1848), a Louisiana physician.

CHOISYA, Mexican Orange-Blossom. Only one of the six species is likely to be found in a British garden, *C. ternata*, and that is most likely to survive on a sunny wall. Jacques Denis Choisy (1799-1859) was a Swiss botanist and a professor at Geneva.

CLARKIA. These hardy annuals from North America are

named after a hardy soldier who saw service against Red Indians under General Wayne. Captain William Clark (1770-1838) was joint leader with Capt. Meriwether Lewis (1774-1809)- of Lewisia fame - of the first trans-continental expedition which crossed the Rockies to the Pacific in 1806-7.

CLIVIA, Kaffir Lily. Lady Charlotte Florentina Clive, Duchess of Northumberland and granddaughter of Clive of India who died in 1868 is supposed to have been the first person in England to grow these African members of the Amaryllis family.

DAHLIA. They are not named after Bertie Wooster's aunt, it is the other way round, and I have never come across one outside fiction. The plants come from the high sandy plains of Mexico and were named after Andreas Dahl (1751-1789), a Swedish botanist and pupil of Linnaeus who tried in vain to prevent his master's papers and collections from being removed to London.

DEUTZIA. There are 50 of these hardy shrubs coming mainly from China, Japan and the Himalayas. They are named after Johan van der Deutz (1743-1788), a Sheriff of Amsterdam and friend and patron of Carl Thunberg, one of Linnaeus's pupils (see below).

DOUGLASIA is not an interesting genus for gardeners but I include it because, as a specific name, David Douglas is remembered in several splendid plants as different as the Douglas Fir and the Poached-Egg Flower (*Limnanthes douglasii*). He was a Scottish collector for the Horticultural Society of London, who went to Oregon and brought back to Europe many things from the Pacific Coast. He died in Hawaii when he fell into a trap for wild cattle and was gored to death by a bull which was already trapped.

ECHEVERIA. These succulents with their rosettes of leaves and loose sprays of flowers are named after the botanical artist who made hundreds of drawings of Mexican plants between 1787 and 1797. He was Athanasio Echeverriay Godoy.

ESCALLONIA. Antonio José Escallon y Flores was an eighteenth century botanist who travelled to S. America where these shrubs come from. He corresponded with the younger Linnaeus.

ESCHSCHOLZIA, Californian Poppy, Noah's Nightcap. Johan Friedrick Eschscholz (1793-1831) was a Russian naturalist who accompanied Otto von Kotzebue on his expedition round the world between 1815 and 1818.

FORSYTHIA. There are European ones but the best of them arrived from China and Korea in the mid-nineteenth century. They were named after the Scot William Forsyth (1737-1804) who worked for the Duke of Northumberland at Syon House and then became Superintendent of the Chelsea Apothecaries' Garden and Royal Gardener to George III at Kensington. He wrote a treatise on *The Management of Fruit Trees*, invented a disgusting concoction called Forsyth's Plaister, which he claimed would heal wounds in the bark of trees, and he is alleged to have made the first rock-garden in England with stones taken from the Tower of London.

FOTHERGILLA, American Wych-Hazel. These relatives of *Hamamelis* have bottle-brush flowers and colour brilliantly in the Autumn. Dr John Fothergill (1712-1780) was a Quaker physician of Stratford in Essex who specialised in growing plants from America.

FREESIA. These lovely scented relatives of the Irises were named after a German physician of Kiel who died in 1876.

FREMONTODENDRON, Flannel Bush, comes from California and Mexico and is named after Major-General John Charles Frémont (1813-1890) who, like David Douglas, made perilous expeditions to the American West Coast and brought back new plants.

FUCHSIA. Leonhard Fuchs (1501-1566) see above.

GAILLARDIA, Blanket Flower. These bright red and yellow daisy-like annuals were named in 1786 after Gaillard de Charentonneau, a French magistrate who was also a patron of botany.

GALTONIA, Summer Hyacinth, is named after Sir Francis Galton (1812-1911), a cousin of Darwin, who was an explorer in his younger days and then made many important contributions to science, among other things, pioneering the study of eugenics and the use of fingerprints in identification.

GARDENIA, Cape Jasmine. There are 250 species but only one in cultivation, the strongly fragrant ivory-flowered *jasminoides*, a favourite for buttonholes. They are called after the happily named Alexander Garden (1730-1791), a physician who emigrated from Aberdeen to Charleston, S. Carolina, and corresponded with Linnaeus.

GERBERA, Transvaal Daisy. These large, showy, orange-scarlet daisies are not fully hardy. They were named after a German naturalist who travelled in Russia, Traugott Gerber, who died in 1743.

GESNERIA This family contains beautifully tender things like Gloxinia and Streptocarpus and was named after an eminent naturalist of Zurich, Conrad von Gessner (1516-1565). He is credited with bringing the tulip into favour in western Europe which was to lead to the extraordinary tulip mania of a century later. He wrote an important *Historiae Animalium*, but was carried off by the plague before he could complete a comparable volume on plants.

GLOXINIA. Benjamin Gloxin was a doctor whose *Botanical Observations* were published at Strasbourg in 1785. However the popular pot-plants with big Foxglove-like flowers commonly known as Gloxinias should properly be called Sinningias after the unhappily named William Sinning (1794-1874) who was head gardener at the University of Bonn.

GODETIA. These showy annuals come from N. America but are named after Charles Godet (1797-1879), a Swiss botanist who wrote a flora of the Jura.

GRISELINIA. These shrubs from New Zealand with bright green leaves made good hedging in mild and seaside districts. Francesco Griselini (1717-1783) was a Venetian naturalist.

GUNNERA. The one you are most likely to see in this country, usually planted close to water, is *G. manicata* which comes from Brazil. My children used to call it Martian Rhubarb. Ernst Gunnerus was a botanising bishop who wrote a flora of Norway. His dates were 1718 to 1783.

HALESIA, Snowdrop Tree, comes from N. America and is named after the ingenious Rev. Stephen Hales (1677-1761) of Teddington. He was a physiologist and chemist who invented, among other things, a way of ventilating prisons and of fumigating grain with brimstone. He succeeded in measuring blood-pressure in a gruesome experiment involving his grey mare, a gate and a goose's windpipe.

HEUCHERA, Coral Flower, Alum Root. *H. sanguinea* makes useful ground-cover under trees and looks particularly well in a vase with other cut flowers. J.H. von Heucher (1677-1747) was Professor of Medicine at Wittenburg.

INCARVILLEA. Father Pierre d'Incarville (1706-1757) was a French missionary in China whence he introduced many plants to Europe, including these pink-flowered perennials with their attractive toothed foliage.

KALMIA, Calico Bush or Mountain Laurel are English names for *K. latifolia* and the other species that is often grown, *K. angustifolia*, is called Sheep Laurel. These lovely shrubs which mix so well with Rhododendrons were named after Pehr Kalm (1715-1779) a Finnish pupil of Linnaeus who was sent to N. America, where they come from, by the Swedish government. He wrote reliable reports on what he discovered about the natural resources of the continent.

KERRIA, Jew's Mallow. There is only one species, called *japonica* because it was much cultivated there, though it came originally from China. It was one of many plants collected for Kew in the Far East by William Kerr who died in 1814.

KNIPHOFIA, Red Hot Poker, Torch Lily. These familiar members of the Lily family come from Africa but were named after Johan Hieronymus Kniphof (1704-1763) who was a Professor of Medicine at Erfurt and produced a volume of illustrations of plants.

KOLKWITZIA, Beauty Bush. The peeling brown bark of this shrub is almost as attractive as the pink, foxglove-like flowers. There is only one species and it comes from W. China. Richard Kolkwitz was born in 1873 and became a Professor of Botany at Berlin early this century.

LAVATERA, mallow. There are annual and biennial members of the genus as well as the slightly tender Tree Mallow. The two Lavater brothers after whom they are named were doctor-naturalists in Zurich during the sixteenth century.

LEWISIA. See under Clarkia above.

LINNAEA, Twin Flower, can be attractive on a shady rockery. With typical modesty and humour Carl von Linné wrote, "A plant of Lapland, lowly, insignificant, disregarded, flowering but for a brief space - from Linnaeus who resembles it." He also named an ugly and smelly weed after one of his enemies!

LOBELIA was introduced from South Africa in 1752 but there are 200 species in the genus, many from N. America. They were named after Mathias de l'Obel who died in 1616, the same year as Shakespeare. He was Flemish but became Royal Botanist to James 1st of England. Tincture of Lobelia was once used in treating asthma.

MAGNOLIA. There are about eighty species and they come from N. America as well as the Himalayas, Japan and China where the buds of one kind are used to flavour rice. They are named in honour of Pierre Magnol (1638-1715), Professor of Botany and Director of the Botanical Gardens at Montpellier.

MAHONIA. These evergreen shrubs with sweetly scented yellow flowers in early Spring are named after Bernard MacMahon (1775-1816) who emigrated from Ireland to Philadelphia, where he established a nursery garden and seed shop which became a meeting place for botanists and horticulturalists. He was author of *The American Gardener's Calendar*. *M. aquifolium* is called Oregon Grape but other species come from Asia and all have blue-black berries.

MONARDA, Bee Balm, Bergamot, Oswego Tea. Nicholas Monardes (1493-1588), physician and herbalist of Seville, published a book about new products from America which was translated into English in 1577 as *Joyful Newes out of the Newe Founde Worlde*.

MONTBRETIA was introduced from the Cape of Good Hope in the early nineteenth century. We now have to call it Crocosmia. It was named after the young botanist who accompanied Napoleon's Egyptian expedition in 1798 and died there at the age of twenty, Antoine-Francois-Ernst Coquebert de Montbret (1781-1801).

NICOTIANA, Tobacco Plant. The word 'tobacco' is supposed to come from a Carib name either for the cigar-like roll of leaves or from the pipe in which the weed was smoked. It was introduced into France by Jean Nicot (1530-1600), the French ambassador to Lisbon.

OLEARIA, Daisy Bush. These evergreens mostly come from New Zealand and are tough as old boots, shrugging off city grime or salt spray indifferently. I think their flowers smell disgusting. The name looks as if it ought to mean 'Olive-like' though *O. ilicifolia* is 'Holly-like' and known as Maori Holly: in fact they are named after Adam Ölschläger (1603-1671) whose name was Latinised as Olearius. He wrote a flora of Halle.

OSMUNDA, Royal Fern. *O. regalis* is hardy and will grow to ten feet in the wild, a splendid sight. The name is of uncertain origin but may be from Asmund or Osmund, a thirteenth century writer of runic verses who paved the way for the Christian conversion of Sweden.

PERNETTYA. There are twenty species but only one, *P. mucronata*, (sharp-tipped) is in cultivation. The berries can vary in colour from white through many shades of pink and red to purple but it is unisexual so you need male and female plants to get any joy from it. Antoine Pernetty (1716-1801) found it in S. America when he was travelling there with Bougainville in the 1760s.

PUSCHKINIA, the Striped Squill, was collected in the Caucasus in 1802 by the Russian Count Mussin-Puschkin.

ROBINIA, False Acacia, Locust. Their flowers tell you these trees from the United States belong to the Pea family. They were named in memory of Jean Robin (1550-1629) who was Royal Gardener to two French kings. He had over a thousand different species in cultivation at the Louvre Palace and exchanged specimens with Gerard and Parkinson in England.

ROMNEYA, Tree Poppy. There are two species, both white-flowered and from California. *R. coulteri* (from the Dr Coulter who discovered it) can grow to a magnificent six feet and more. The genus is named after Thomas Romney Robinson (1792-1882), an Irish astronomer.

ROSCOEA. These natives of China have roots like a Dahlia and flowers like an Orchid which are attractive in a rock-garden. William Roscoe (1753-1831) of Liverpool was founder of the botanical gardens there, an anti slave-trader and author of a splendid book about the Ginger family.

RUDBECKIA, Cone-Flower, Black-eyed Susan. These hardy perennials of the herbaceous border were named by Linnaeus after

Olof Rudbeck the Elder (1630-1702) who founded the Botanical Gardens at Uppsala. The wood-blocks illustrating 11,000 flowers on which he spent much of his life were destroyed in a fire in 1702, the year of his death.

SAINTPAULIA, African Violet. *S. ionanthe* (violet-flowered), the popular pot-plant, was discovered in E. Africa by the German Baron Walter von Saint Paul-Illaire (1860-1910). There are eleven other species.

SANSEVIERIA, Mother-in-law's Tongue. There are sixty species but *S. trifasciata* (in three bundles, from the arrangement of the flower-clusters) is the familiar pot-plant with upright sword-like leaves. The genus is named after Raimond de Sangro, Prince of Sanseviero, who was a patron of horticulture in the eighteenth century.

SCHLUMBERGERA, Christmas Cactus. These epiphytic cacti, which seem as much at home in our living-rooms as in the jungles of Brazil were named after a Belgian explorer and plant collector, Frederick Schlumberger (1804-1865).

SINNINGIA, see under Gloxinia.

SMITHIANTHUS, Temple Bells. These handsome greenhouse flowers from Mexico were named after Matilda Smith (1854-1926), an illustrator for the Botanical Magazine.

STERNBERGIA. These bulbous plants are sometimes called Winter Daffodil though the flower is more Crocus-like apart from appearing on a true stalk. Count Kaspar von Sternberg (1761-1838) was an Austrian botanist who founded the Bohemian National Museum at Prague.

STRELITZIA, Bird of Paradise Flower. Many plane-loads of pink holiday-makers returning from the Canary Islands have clutched bunches of these spectacular flowers which need greenhouse treatment here. George III's unhappy Queen Charlotte was a Mecklenberg-Strelitz and they are named for her.

STUARTIA. These shrubs have attractive bark, autumn foliage and red seed-capsules as well as white saucer-shaped flowers. Is any other plant named after a Prime Minister? John Stuart, 3rd Earl of Bute (1713-1792) was British Premier briefly in the 1770s. Linnaeus mistakenly called it Stewartia and so it should strictly remain.

TOLMIEA, Piggy-back Plant. Often grown as a pot plant because of its curious way of producing young plants on the leaves, this Saxifrage is quite hardy and can be used for ground-cover. Dr W.F. Tolmie (1830-1886) was a Scottish physician, botanist, ethnologist and fur-trader in N.W. America.

TORENIA, Wishbone flower. *T. fournieri* has become quite popular as a pot-plant recently for its velvety purple-lilac flowers. It was named after the Rev. Olof Toren (1718-1753) who was chaplain to the Swedish East India Company at Surat and in China.

TRADESCANTIA, Spiderwort, Wandering Jew. There were two John Tradescants, father and son, who were Royal Gardeners to both King Charles I and II and introduced many new plants to England, including Cos Lettuce, the Larch, Lupins, Michaelmas Daisies, Bergamot and Virginia Creeper. Spiderwort was named after John the Elder (1608-1662) who visited Virginia in 1654.

WEIGELA. These summer-flowering hardy shrubs from China are so popular that it is surprising they have not acquired an English name. Christian Weigel (1748-1831), after whom they are named, was a botanist and Professor at Griefswald in North Germany.

ZANTEDESCHIA, Arum Lily, Calla. The cloying scent of white Arum Lilies is indissolubly associated for me with the cool interiors of Roman Catholic churches in France and Italy. They come from Southern Africa but take their name from Francesco Zantedeschi, an Italian botanist who was born in 1797

ZINNIA, Youth-and-Old-Age. These bright annuals come from Mexico. Linnaeus named them after the German botanist Johann Gottfried Zinn (1727-1759).

Chapter 5

An Anthology Of Annuals

In evolutionary terms, annuals appear on the scene much later than trees and other perennial plants. They are, of course, entirely dependent on setting viable seed for their survival. Many genera now contain annual, biennial and perennial species so the name of a plant which you grow as an annual, Larkspur for instance which is an annual Delphinium, may have been dealt with in another chapter.

Garden writers usually group annuals and biennials together for convenience and I follow their example: the line between them is not a hard and fast one. In their native countries many plants we think of as annuals actually spring up at the end of one season and flower the next. Autumn is, after all, nature's time for dispersing and sowing seeds. Similarly, some plants we treat as biennials, like the Wallflowers, are actually perennials and once established on an old wall will grow larger and woodier for many years. Some true annuals can grow remarkably woody in one season, (think of the Walking-stick Cabbage), or surprisingly tall, (a ten-foot Sunflower is no record). Genuine hardy biennials such as Honesty, Foxglove or Evening Primrose seed themselves so readily that they seem, like the poor (and the weeds), perennially with us. Indeed, many of the half-hardy annuals in this chapter are actually perennial in the warmer countries of their origin. Antirrhinums from the Mediterranean will not stand our winters though they do seed themselves: I always seem to have dark red ones uninvited in my onion bed: but they barely reach the flowering stage by the time of the first frosts and that is why we have to treat them as half-hardy annuals though, at home, they would grow into bigger bushes for several years. When some of the South American Verbenas were introduced last century, Victorian gardeners were advised to bury a small hurdle or the lid of a hamper under the plot where they were to be grown so that they could be lifted without disturbance into the greenhouse for the winter. We avoid this performance by raising them, and many other exotics, afresh from seed each year in heat. And heat, today, is expensive which is undoubtedly one of the reasons why displays of summer bedding are somewhat less common in public and private gardens than they once were. Hardy annuals sown out of doors will not provide colour until after mid-summer and raising tender ones is labour-intensive and labour, too, is expensive. Both hardy and tender ones usually need dead-heading and annuals which will flower for a long time without attention such as Ageratum or Tagetes are always likely to be popular. Even where ornamental gardens are given over to permanent planting- shrubs, bulbs and ground cover - it is worth finding space among the vegetables for a few rows of annual flowers for cutting and many can be seen (and smelt) best in all their beauty as pot-plants.

It is a constant surprise to me that many beautiful and popular flowers - Cineraria, Nemesia, Petunia, Salpiglossis - do not seem to have attracted any common popular names. At least the scientific names become more interesting and memorable when you know what they mean.

AGERATUM, Floss Flower. There are 60 species mostly from C. America and the name means 'unageing' from Greek *a geras*, literally 'not old'. The one most commonly grown for long-lasting summer bedding is the small *houstonianum* named after a Scottish plant collector, William Houston (1695-1733), and there are pink, white and mauve forms as well as the familiar powder-blue one.

AGROSTEMMA githago (*Lychnis* in older books). Corn Cockle was once a serious cornfield nuisance because its toxic seeds could not be separated from the grain by winnowing and spoilt bread and malt. It is becoming a rarity in the wild but makes a very satisfactorily resilient garden plant, especially the 'Milas' variety named after the Turkish town where it was found. *Agrostemma*, Linnaeus' name, means 'field-garland' and *githago* is the old Latin name (green with red-purple stripes). 'Cockle', *coccel* in Old English, looks as if it comes from Latin *coccum*, scarlet, but names for the plant in other European languages suggest that it may primarily be associated with the cock's comb, like the Corn Poppy which is *coquelicot* in French.

ALCEA, Hollyhock. Until recently they were called *Althea*,

Theophrastus' name for the Marsh Mallow from Greek *althaine*, to heal. This was the medieval 'holy Hocc' or Mallow because of its medicinal virtues. *Alcea* is from *alkaia*, the Greek name for another Mallow. *A. rosea* was the first to be introduced, from China; it had only decorative use (though the leaves are edible) but was given the old name for the Marsh mallow by Turner. Another Hollyhock, *A. ficifolia* (fig-leaved), came much later, from Siberia, and is somewhat more resistant to the deadly rust.

ALONSOA, Mask Flower. In their native Peru they are perennial but treated as half-hardy annuals here. They were named after Alonso Zanoni, an eighteenth century Spanish Secretary of State for Columbia at Bogota.

ALTHEA, see Alcea.

AMARANTHUS, Love Lies Bleeding, Tassel Flower. The name means 'not withering' and the Greeks probably applied it to species of *Helichrysum* or *Gomphrena* (Globe Amaranth). When poets like Milton and Tennyson use the name it is a symbol of immortality and it is unfortunate that none of the plants left in the genus today is an everlasting. Some are Spinach-like pot-herbs and the best known of the decorative ones is Love Lies Bleeding, *A. caudatus* (with a tail), which has the even more gory name of 'Nun's Scourge'. Two others have been grown since Tudor times, Joseph's Coat, *A. tricolor* with gold, red and green foliage, and Prince's Feather, *A. hypochondriacus* (melancholy, sombre), which is actually less drooping than *A. caudatus*.

AMMOBIUM, Everlasting Sandflower. These Daisy-like Australian desert flowers get their name from Greek *ammos*, sand, and *bio*, to live.

ANTIRRHINUM, Snapdragon, Lion's Mouth. There are about 40 species: the familiar *A. majus* is a perennial in S. Europe where it comes from. It was known to the Tudor herbalists as Calve's Snout but they do not seem to have found any uses for it. The Greek name means 'like a snout or muzzle': I suppose there are deprived children who have never had the small pleasure of making it open by pinching the sides of the corolla.

ARCTOTIS, African Daisy. There are many species but the

most colourful and long-lasting garden flowers are hybrid perennials grown as annuals. The name means 'bear's ear' from a supposed resemblance of the scales of the pappus (the feathery hairs on the seeds - from Greek *pappos*, a grandfather).

ARGEMONE, Prickly Poppy, Devil's Fig. The English names belong to *A. mexicana* which has scented yellow or orange flowers and striking white-veined thistle-like foliage: *A. grandiflora* has larger white flowers much like a Romneya and there are other species in cultivation. The name was used by Dioscorides for a Poppy with sticky yellow sap which was supposed to cure cataract of the eye, *argemon* in Greek.

ATRIPLEX, Orache, Salt Bush, Shrubby Goosefoot. Orache, *A. hortensis* (of gardens), was once much used like a Spinach but its purplish flower spikes and dark foliage make it a handsome and unusual plant for the flower border, especially the variety *Rubra* with crimson leaves and violet stems. '*Atriplex*' is Pliny's name (perhaps containing Greek *atri*, very dark) and 'Orache' comes from the same source by way of French and Italian.

BRACHYCOME means 'short hair' from the pappus bristles. *B. iberidifolia* (with leaves like *Iberis*, Candytuft, is the Swan River Daisy from Australia and the only species commonly grown, 'Purple Splendour' being the most striking variety.

CALENDULA officinalis. The Pot Marigold, unlike our useless native 'Golds', has long been highly regarded: even the look of the flowers and their pungent aroma were once thought to be health giving. The petals were used as a poor-man's Saffron for flavour and colour in cooking and even as a hair dye. It was introduced in the Middle Ages from the Mediterranean area where it can be found in flower all the year round and that is presumably how it got associated with the calendar. Latin *calendae* were the days fixed, usually the first of the month, for settling debts.

CALLISTEPHUS, China Aster. Seeds of the only species, *chinensis*, reached Europe early in the 18th century and soon many varieties had been raised. The sight of them is said to have inspired Pierre d'Incarville, the great Jesuit plant collector, to go to China. The name combines Greek *kallos*, beautiful, and *stephanos*, a crown.

CELOSIA. Most commonly grown of these tender relatives of Amaranth from tropical Asia is *C. argentia* (silver (flowered)), though nearly always in cultivated forms with garish bright scarlet and yellow colours which justify the generic name from Greek *kelos*, burning. These are *C.a. cristata* (crested), known as Cock's Comb and formerly as Globe Amaranth or Floramor, and *C.a. plumosa* (feathery), also known as *pyramidalis*.

CENTRANTHUS means 'spur-flower': it is also spelt *Kentranthus* (from Greek *kentron*). Red Valerian, *C. ruber*, which has escaped on to cliffs in the south of England, is a perennial but *C. macrosiphon* (large tubed) is a useful rosy annual with a white form.

CHEIRANTHUS. The Wallflower is probably not a native but has long been established here and, when happy, is not only perennial but surprisingly woody and long-lived. Nevertheless, for bedding it is always treated as a biennial. The name probably comes from Greek *cheir*, hand, and *anthos*, flower, suggesting that it was carried in bouquets, though some authorities think the first element is Arabic. It has been known in England by a confusing number names, Sweet William, Chevisaunce (meaning 'comfort') and yellow Stock Gilliflower - the association with Stock and Carnation because of the powerful scent. Bacon wisely said, "It is very delightful to be set under a chamber window". The orange Siberian Wallflower has a distinctive scent and combines wonderfully with Forget-me-not. It is strictly an *Erysimum* (see p.121) but still usually listed as *C. allionii*, after Carlo Allioni, an Italian botanist who wrote a flora of Piedmont.

CINERARIA cruenta (blood-red), from the Canary Isles, is the parent of the popular spring-flowering pot plants but the nearly hardy *C. maritima* is grown as an annual for its attractively divided silver-grey leaves. It is now properly classed as *Senecio bicolor*: this is a pity since *cinereus* means ash-grey from the down on the foliage while *Senecio* suggests the fluffy white hair of an old man from the seeds.

CLEOME, Spider Flower. The only one at all commonly grown is the striking tropical American *C. spinosa* (spiny), also known as *C. pungens* with the same meaning of 'pointed' (nothing to do with being scented, which it is). The origin of the generic name seems uncertain; Greek *kleos*, glory, seems more likely to be involved than *kleio*, I shut.

COLLINSIA bicolor, from California, has lilac and white or, less commonly, pink flowers and, unlike most annuals, is happy with partial shade and damp soil. It was named for the Vice President of the Philadelphia Academy of Sciences, Zaccheus Collins (1764-1831) and his surname, incidentally, is a double diminutive of Nicholas so Nixon and Collins are essentially the same name.

CONVOLVULUS comes from the Latin verb *convolvo*, I twine about, a name used by Pliny, but the European annual *C. tricolor* (or *C. minor*) is not a twiner, but a neat bedder whose purple, mauve or pink funnel-shaped flowers have white and yellow centres. It is sometimes called Morning Glory because the individual flowers, though profusely borne, only last one day. However, that name is more often associated with the tropical *Ipomoeas* who certainly do twine and whose name means 'worm-like'.

COSMOS from the Greek *kosmos* (beautiful) does not seem to have attracted any English names except Mexican Aster, though two species of half-hardy annuals are deservedly popular for producing bright-coloured single Dahlia-like flowers over a long period. These are *C. bipinnatus*, with feathery (bipinnate) leaves and a colour range similar to *Pyrethrums*, and *C. sulphureus* with sulphur-yellow or orange flowers.

COTULA gets its name from the way the leaves make a small cup (*kotula*) where they join the stem. *C. barbata* (bearded - the stem), the Pincushion Plant, is a neat little annual perhaps best suited to the rockery. If you grow it as a pot plant, you will find it has a strong and rather disagreeable scent. It comes from S. Africa and there are other species from that country and New Zealand.

CUPHEA gets its name from Greek *kuphos*, curved, from the shape of the seed pod. They come from C. and S. America and two species can be grown as half hardy annuals (or as pot plants), the Cigar Flower (*C. ignea* = glowing) with a cigar-shaped corolla and *C. miniatus* (orange-red like cinnabar or red lead).

DIASCIA, Twinspur, from southern Africa has pretty shell-like flowers with two spurs, *di askos*, literally 'two sacs' in Greek. There are mat-forming species for the rockery and the larger *D. barberae* which has pink flowers with a purple eye: it was named after a plant collector, Mary Elizabeth Barber née Bowker (1818-1899).

DIDISCUS (two discs), see Trachemene.

DIGITALIS. The Fuchs who is commemorated by the Fuchsias gave the genus its name from the German *Fingerhut*, a thimble: *digitus* is finger in Latin. The old herbalists like Fuchs ascribed several fanciful virtues to the plant but the discovery of digitalin for treating heart disease belongs to the very end of the 18th century. Grigson says the name 'Foxglove' originally belonged to Deadly Nightshade, *Atropa belladonna*, but it is known as 'Fox-bells' in some other languages and it has been suggested that the English name was not originally -glove but *gleo(w)*, an old musical instrument of graduated bells. There are many other country names such as Bee Catchers, Deadman's Bells and Pop-Dock (the unopened flowers can be popped). For the garden there are introduced species like the yellow *D. lutea* as well as showy varieties of our native *D. purpurea* though the naturally occurring white form is, for my money, the most graceful of all for a woodland setting and quite magical in the twilight.

DIMORPHOTHECA. Linnaeus called these sun-loving African Daisies *Calendulapluvialis*, the Rain Marigold, because they shut when the sky darkens. They are now often listed as *Osteospermums* (bone-seed, from the shape) but the present name, which is even more of a mouthful, means 'two shaped container' because the fruits produced by the ray flowers are different from those of the disc. The most popular ones today probably result from crossing *D. aurantica* (orange-coloured) with *D. sinuata* (wavy, the leaf-edge) and are known as Star of the Veldt, but there are several other species giving a range of colours, the upper and under sides of the petals often being strikingly different shades.

DOROTHEANTHUS, see *Mesembryanthemum*.

EUCHARIDIUM means 'full of grace, pleasing'. *E. concinnum* (neat, elegant), an annual from California called 'Red

Ribbons' with petals like a Ragged Robin is now usually classified as *Clarkia concinna*.

FELICIA. Best known is the Kingfisher Daisy, *F. bergeriana*, named after the German botanist Alwyn Berger (1871-1931). It is something like a mat-forming Michaelmas Daisy, having blue flowers with gold discs. The generic name is variously said to be after a German official called Felix or directly from Latin *felicitas*, cheerfulness.

GAZANIA, Treasure Flower, is one of the finest of the many African Daisies which have to be treated as annuals here. Usually grown today are hybrids bred from *G. rigens* (rigid), also known as *splendens* which is bronze-orange with dark central markings. The dark green leaves have silky undersides. Theodore Gaza (1398-1478), after whom they are named, translated Theophrastus's works into Latin.

GODETIAs from N. America, relatives of the Evening Primrose, are now often included with the Clarkias. There are single and double or 'Azalea-flowered' varieties of *G. grandiflora*, the most popular species, in many shades of pink and red. C.H. Godet (1797-1879) was a Swiss botanist who compiled a flora of the Jura.

GOMPHRENA globosa. The Globe Amaranth, originally from India, may well be the immortal Amaranth of the ancient Greeks. Its everlasting Clover-like flower head was thought to resemble a club, *gomphos* in Greek, and they are usually in shades of red, purple or pink.

HELICHRYSUM, Straw Flower, Immortelle. The name of this big genus means 'sun-gold'. The familiar many-coloured annuals from Australia with their papery everlasting flowers are varieties of *H. braceatum* (having (conspicuous) bracts). *H. orientale* from Crete and Asia Minor has a much longer history in cultivation and was being used for the winter decoration of churches in the 17th century.

HELIOPHILA longifolia has four-petalled blue flowers with a white eye and came from S. Africa. The name means 'sun-lover'.

HELIPTERUM is made up of 'sun' and 'wing' (*pteron*) because they are sun-loving everlastings and their seeds have feathery wing-like hairs. Among the annuals there are, besides the pink *roseum*, *H. humboldtianum* with fragrant yellow flowers named after Alexander von Humboldt (1769-1859) the great German naturalist-explorer, and the red and white *H. manglesii* after Robert Mangles who introduced and grew many plants from W. Australia, the home of the genus, in the 19th century.

IMPATIENS, Busy Lizzie, Touch-me-not, Jumping Betty. Our yellow native Balsam, *I. noli-me-tangere*, has seeds which, as Parkinson puts it, "will soone skippe out of the vessels when ripe", and this family characteristic gives rise to the scientific name meaning 'impatience' and to most of the common names. The six-foot Himalayan Balsam, *I. glandulifera* (having glands), which is now naturalised by many of our waterways is sometimes known as Policeman's Helmet. It is a big genus but most of the pot- and bedding-plants sold as Busy Lizzies today are hybrids of *I. sultanii*, named by Sir Joseph Hooker after the Sultan of Zanzibar, whence it came, and *I. walleriana* after Horace Waller, a 19th century missionary in central Africa.

KENTRANTHUS, see Centranthus.

KOCHIA, Burning Bush, Summer Cypress, is named after a German professor of Botany, Wilhelm Koch (1771-1849). *K. scoparia* (broom-like) was introduced from S. Europe in the17th century and fitted well into the formal planting of the time. It was taken to be a Linaria and so called Broom Toad Flax. Usually grown today is the variety *K.s. Trichophylla* meaning 'hair-like leaves' which turn scarlet in autumn and give rise to its commoner popular name.

LAYIA elegans, Tidy Tips, is one of a number of bright, hardy, Daisy-like annuals from California. The genus was named after the botanist attached to an expedition to the Pacific N.W. and Bering Straits, G. Tradescant Lay, who died in 1845.

LIMNANTHES douglasii, Meadow Foam, Poached Egg Flower. The name means 'marsh-flower' reminding us that, though it was collected by the great David Douglas (see *Douglasia* p.37) in sunny California, it still needs plenty of moisture. The yellow and white flowers are very attractive to bees.

LIMONIUM. Our native Sea Lavender, *L. vulgare*, used to

be called Statice, Dioscorides' name meaning 'astringent'. It grows in muddy salt-marshes. 'Limonium' is another of Dioscorides' names, said by the authorities to be for a meadow flower from Greek *leimon*, a meadow. However, in Latin, *limosus* means 'muddy'. The species most commonly grown in gardens with everlasting rose, white, yellow or mauve flowers is *L. sinuatum* (from the wavy-edged leaves), but there are other useful species and perennials as well as annuals.

LINUM is the Latin name for Flax (Greek *linon*). The ancient Egyptians wrapped their mummies in linen and Flax, *L. usitatissimum* (the most useful), provided the most important vegetable fibre (not to mention the many uses of linseed oil) until the introduction of cotton in the 18th century. Our native *L. anglicum* is also blue-flowered but a perennial and there are yellow-flowered perennials for the garden like *L. flavum* from Russia, but showiest is the annual Scarlet Flax, *L. grandiflorum*, from N. Africa.

LUNARIA is from Latin *luna*, moon, and 'Moonwort' is one of its English names from the round, papery seed-pods which have inspired many other names like Satin Flower, Shillings and the French *Herbe aux lunettes*, 'Spectacle Plant'. Gerard says, "among our women it is called Honestie", presumably because you can see precisely how many pennies it has in its transparent purse, and this is still the commonest name. It is biennial, (*L. biennis*), but seeds itself so freely even in the bottom of hedges that it is likely to provide a perennial patch of purple in Spring when that colour is scarce.

MALCOLMIA maritima, the sweetly scented Dwarf Annual Stock, got the name 'Virginian' by mistake as it is a native of the Mediterranean area. The French call it *Giroflee de Mahon* after the port of Minorca. It was originally classed as a *Cheiranthus* but renamed in honour of two William Malcolms, father and son, who were Kensington nurserymen in the late 18th and early 19th centuries.

MALOPE trifida is an annual Mallow, the 'Grandiflora' variety being particularly handsome and free-flowering. The name was used by Pliny for a kind of Mallow and, like *Malva*, is presumably connected with Greek *malos* meaning 'soft' or 'tender' and referring to the leaves which, in this species, are also three-lobed (*trifida*).

MATRICARIA exima (excellent), Feverfew, Mayweed, is now properly *Chrysanthemum parthenium* (chaste, virginal) and there are attractive garden varieties grown as annuals. '*Aureum*' has golden green foliage which lasts through most winters. *Matricaria* means 'mother-care', the plant having been used medicinally for problems of the womb (Latin *matrix*).

MATTHIOLA, Stock. The Night-Scented Stock, *M. bicornis* (two-horned, the fruit) comes from Greece and is a straggly little thing best sown together with the Virginian Stock (see *Malcolmia*) for colour by day and scent at night. Our other garden Stocks have two natives as their parents, *M. sinuata*, the Sea Stock of our S.W. shores and, particularly, *M. incana* (grey-white), the Hoary Stock which is really at home in S. Europe but has a toehold in Britain on the cliffs of the Isle of Wight. The Ten Week Stock used to be called *M. annua* but is now thought to be only a variety of *incana*, actually a short-lived perennial. Brompton Stocks were raised at a nursery in that part of London. Stocks were originally called Stock Gilliflowers because they had a woody stem or stock and the generic name comes from a 16th century Italian, Pierandrea Matthioli, who was physician to the Emperor Maximilian II and wrote a much-translated commentary on Dioscorides.

MENTZELIA lindleyi, Prairie Lily, Blazing Star, is a hardy annual from California with golden flowers somewhat like a *Hypericum*. It used to be called *Bartonia aurea* after F.R. Barton (1865-1947), an Army Captain and colonial administrator, but now commemorates two other botanists; Christian Mentzel (1622-1701) was physician to the Elector of Brandenburg and John Lindley (1799-1865) was an English professor.

MESEMBRYANTHEMUM. The name has an unusual history. It was originally coined in the 17th century from Greek *mesembria*, 'midday-flower', but when night-flowering species were found in the next century, Dillenius made a small change to the spelling (i to y) so that it now refers to the central position of the ovary (*embryon*). We shall have to unlearn the name anyway, since the big genus of 1000 species is now being split into several smaller ones and the popular Livingstone Daisy, *M. criniflora* (hair-flower) is now being called *Dorothenathus* after the mother of the botanist Dr Martin Schwantes and *bellidiformis*, 'Daisy-like'. It seems that the explorer, David Livingstone, has less to do with the name than the fact that some pebble-like members of the genus are like 'living stones' (*lithops*). The annual Ice Plant, *M. crystallinum* (from the frosty crystalline appearance of the leaves) has been in cultivation much longer and is now to be called *D. tricolor*.

MIRABILIS. The Latin name means 'wonderful' and *M. jalapa* (from Xalapa in Mexico) is known as Marvel of Peru or Four o'clock Plant from the afternoon opening of the flowers. There are other species also grown as half-hardy annuals but their popularity has been somewhat eclipsed by later introductions like *Petunias*.

MOLUCCELLA laevis (smooth), Bells of Ireland, Shell Flower, Molucca Balm. The name is presumed to have come from the Moluccas, formerly the Spice Islands, in the Malay archipelago. The flowers are tiny, white and scented, each being surrounded by a much more conspicuous green shell-like calyx giving the impression of a green flower-spike (which can be dried and used for winter decoration). Whether there is any reason other than the colour to link them to the Emerald Isle, I have been unable to discover.

MYOSOTIS, Forget-me-not. Both the English and scientific names are surrounded by confusions. The names 'Remember-me' and 'Forget-me-not' go back to the 14th century and seem to have been used for the Speedwells and probably other blue flowers as well as the Water Forget-me-not which is clearly pictured by Brunfels in 1543 and called 'Vergissz mein nix'. The Germans had a soppy story about a knight who tried to pick this flower as a keepsake for his lover and cried 'Forget me not!' as the river swept him to his death. Coleridge probably heard the story in Germany, saw the flower commonly planted on graves and popularised the name in England with his poem *The Keepsake* composed in 1802. The Water Forget-me-not is *M. scorpoides* because, as with other species, the flower spike is curled like a scorpion's tail: hence the older name of Scorpion Grass and the belief that it was a remedy for the sting. '*Myosotis*' means 'mouse-ear', a name used by Dioscorides for some other plant for there is no obvious resemblance between any part of the Forget-me-not and that portion of a mouse's anatomy. Garden varieties are usually hybrids of two native species, the Wood Forget-me-not, *M. sylvatica*, and the Alpine *M. alpestris*.

NEMESIA. I know of no English name for this popular summer bedding plant from S. Africa. Dioscorides used the name *nemesion* for a Snapdragon-like plant but its etymology is unknown: Greek *nema*, a thread, or *nemos*, a glade, might be involved. The species usually grown is *N. stumosa* meaning a cushion-like swelling.

NEMOPHILA is one of the Californian annuals introduced by David Douglas in 1833. Some species were found growing in woodland, hence the name which means 'glade-loving'. Even *N. insignis* (striking), Baby Blue-eyes, with its white-centred flowers will stand some shade.

NICANDRA, Shoo-fly Plant, Apple of Peru, is named after Nikander of Colophon, a Greek physician and poet who lived in the second century B.C. There is only one species, *physaloides*, which means 'like Physalis', the Cape Gooseberry (see p.178), the resemblance being that the white-throated lavender bell-shaped flower is followed by a large fruit in a striking purple calyx. Don't try eating it, however; this is a member of the Deadly Nightshade family, *Solanaceae*, and it is said to repel flies.

NIGELLA, Love-in-the-Mist. The genus gets its name from Latin *niger* because the seeds are black like shiny knobs of anthracite. *N. sativa* (cultivated), the Nutmeg Flower, was grown first for its aromatic seeds which were called 'black cummin' and are still used for flavouring in some countries. *N. damascena* (from Damascus) is our usual garden species though *N. hispanica* (from Spain) is a larger plant with darker blue flowers but less hardy. Their inflated seed pods and thread-like bracts have inspired many popular names such as Fennel Flower, Devil-in-a-Bush, Jack-in-Prison, Love-in-a-Puzzle and in France (where else?) *Cheveux de Vénus*, meaning pubic hair.

NOLANA, Chilean Bellflower, gets its name from Latin *nola*, a small bell. The species usually grown is *N. acuminata* (long-pointed) with lavender flowers.

OENOTHERA biennis, Evening Primrose, was introduced

from N. America about 1620 and at first known as the Primrose Tree of Virginia. It has become naturalised in sand dunes and waste places and has a large edible root which enables it to survive in dry places. Several other species have been introduced which are neither night-flowering, yellow nor biennial. They are closely related to Godetias and belong to the Willowherb family. Linnaeus chose the name, one used by Theophrastus for a plant, perhaps an Oleander, the root of which 'gives off a fragrance of wine', as he says, and was thought to make men desire to drink it. *Oinos* is wine and *thera*, imbibing. However, one authority gives the name as meaning 'Ass-catcher'; *ono* is donkey in Greek; so don't always believe what you read in books (including this one!).

ONOPORDUM. These relatives of the Scottish Thistle, *O. acanthium* (spiny), are mainly biennials and *O. arabicum* (Arabian) with silver stems and leaves will grow 8 feet tall. The generic name means 'donkey-fart' from the flatulence they are supposed to cause these thistle-loving beasts.

OSTEOSPERMUM, see *Dimorphotheca*.

PERILLA frutescens (shrubby). The flowers are insignificant though spicy but the red-purple foliage gives the name Beefsteak Plant. The origin of the generic name is said to be unknown, but Perillus was an Athenian sculptor in metal and some species have a bronze sheen to the leaves.

PHACELIA comes from Greek *phakelos*, a bundle, because the clear blue flowers are borne in a cluster. *P. campanularia* is a hairy little annual from California whose flowers are attractive to bees and, to my mind, more like a Speedwell's than a Campanula's. There are larger species which are less often grown.

PLATYSTEMON means 'broad stamen'. There is only one species, the hardy and easily grown *californicus*, known as Cream Cups, with pale Buttercup-like flowers.

PORTULACA was a name used by Pliny meaning 'milk-carrier'. Today it is a large genus but only *P. grandiflora* from Brazil seems to be commonly grown. Its spreading habit makes it suitable for the rock garden. The leaves are cylindrical in section, the stems red and the bright saucer-shaped flowers have a prominent boss of golden stamens which give rise to the name 'Sun Plant'. 'Rose-Moss' is a less obvious name.

RESEDA odorata came from N. Africa and was first known as the Egyptian Bastard Rocket. We now know it by its French name, Mignonette, which might be translated as 'little pet' or 'sweetie'. It was once grown in window boxes to counteract the offensive odours of the street. 'Reseda' is Pliny's name from Latin *resedo*, I heal, and it was once used to soothe bruises. Our sadly scentless native Wild Mignonette is *R. lutea* (yellow) and *R. luteola* (yellowish) is Weld or Dyer's Rocket which gives a strong yellow dye.

RICINUS, Castor Oil Plant, Palma Christi, from tropical Africa is grown for the contribution its palmate leaves make to formal bedding. They can be bright green or purple-bronze though there is only one species, *R. communis* (common). The generic name is Latin for a tick from the appearance of the beans (c.f. *Coreopsis* p.68) which are poisonous though the oil expressed from them has many uses besides lubricating castors and indolent bowels.

SALPIGLOSSIS from Greek means 'trumpet-tongued'; the style is actually clarinet-shaped to my eye. So far as I know *S. sinuata* (from the wavy-edged narrow leaves) is the only species in cultivation but the Grandiflora variety produces a gorgeous range of rich colours each inlaid, as it were, with gold filigree.

SCHIZANTHUS, Poor Man's Orchid, Butterfly Flower. *S. pinnatus* (with leaflets growing opposite each other in pairs) from Chile has produced the varieties usually grown. They might more often be used for summer bedding though, as pot plants, the beauty of the individual flowers is best seen. Each corolla is deeply divided, hence the name meaning 'split-flower'; they deserve a prettier one.

SILYBUM, Our Lady's Milk Thistle. Dioscorides used the name *silybon* for a thistle-like plant and the English names refer to *S. marianum* (of the Virgin Mary) which has foliage handsomely splashed with milky white markings.

SPECULARIA speculum-veneris, Venus' Looking Glass. Latin *specularis* means shining, mirror-like; the names are more than usually fanciful. This little annual may now be classified as a *Legousia* and has white-eyed violet-blue flowers.

TAGETES. African and 'French' Marigolds are misleadingly named since the varieties grown today have been developed from species native to Mexico which became naturalised on the Algerian coast in the 16th century. Their smell was for long thought to be injurious to health. Tages, grandson of Jupiter, sprang from the ploughed earth, a boy with the wisdom of an old man, and was an Etruscan deity reputed to have founded their craft of divination.

TITHONIA rotundifolia (round-leaved), the Mexican Sunflower, is not unlike an orange-scarlet *Zinnia*. A decrepit old man is proverbially a 'Tithonus'. The legend is that he was a beautiful young King of Troy much loved by Aurora who asked the gods to make him immortal but neglected to ask for eternal youth too.

TRACHYMENE caeruleus, Blue Lace Flower, was formerly known as a Didiscus and comes from Australia. It is an *Umbellifer* whose flowers resemble a Field Scabious. The whole plant is sticky to the touch but the name means literally 'rough membrane' and refers to the fruit.

URSINA. There are several annual species of these S. African Daisies: the one most often seen is *U. anethoides* (like *Anethum*, one of the *Umbelliferae*) whose orange flowers have a purple centre. Johannes Ursinus of Regensburg (whose surname meant bear-like) lived in the 17th century and wrote a book about the trees in the Bible.

VENIDIUM whose name means 'veined' from the ribbed fruit is yet another genus of S. African Daisies. *V. fastuosum* (proud) is known as Monarch of the Veldt or Namaqualand Daisy and it has been crossed with an *Arctotis* to produce a valuable hybrid.

VERBASCUM, Mullein. We have several conspicuous native biennials which deserve a place in the garden, notably *V. thapsus* (from the Isle of Thapsos), the Common or Great Mullein which can grow to three or four feet tall. Its woolly leaves give it many country names like Beggar's Blanket and Cuddy-Lugs (i.e. donkey-ears). 'Mullein' comes via Old French *moleine* from Latin *mollis*, soft, and 'Verbascum', Pliny's name, is a corruption of *barbascum*, shaggy or bearded.

VERBENA. Our one native is Vervain (*V. officinalis*) which has a long history in magic and medicine, having been used against wheezings, worms and witchcraft. However, our garden hybrids with their searing colours and strong scent have all been bred from S. American species which, though strictly perennial, are always grown annually from seed. I have seen 'Verbena' explained as having come by way of 'Herbena' from *herba bona*, the 'good plant'. To the Romans, however, *verbenae* were branches of various trees such as Myrtle and Olive used in religious rites and that is the source of the Linnaean name.

VISCARIA elegans, a useful little annual of the Campion family, is now likely to be listed as *Lychnis* or *Silene* with the specific name *coeli-rosa* (rose of heaven) and in the variety *Oculata* (with an eye). *'Viscaria'* means sticky, the Catchflys being members of the genus.

Elecampane
Inula helenium from Leonhart Fuchs' herbal of 1542.

Chapter 6
A Patchwork Of Perennials

The traditional border of hardy perennials is like a patchwork quilt in having, within a rectangle, irregular areas of different colour cunningly placed to complement each other, but it has many more than the three dimensions of length, breadth and colour. There are the different heights of plants to consider and the infinite variety of flower and leaf form and habit of growth as well as the time dimension of what will be out with what when. Many are adaptable but some will have preferences for full sun or some shade, for soil that is rich or poor, moist or dry, alkaline or acid. Planning a border is not an enterprise to be lightly undertaken. It has been simplified in one sense but complicated in another by new species and by the very many new varieties of old ones which have extended the colour range, flowering season and produced both taller, and shorter and more compact, forms of familiar favourites. We are spoilt for choice, embarrassed by riches.

There are inspiring examples of perennial borders at Arley Hall in Cheshire, Packwood in Warwickshire, Hidcote in Gloucestershire, Powys Castle and many other great gardens, but whether we should be inspired *to emulation* is questionable. They may be a peculiarly English invention, associated with William Robinson and Gertrude Jekyll, but they belong to a time when labour was cheap and space more plentiful than it is for ordinary gardeners today. Most of us do not have the high wall or clipped Yew hedge to back a border and, being limited to a rectangular plot, are anxious to avoid straight lines within it. Perennials can be very effective in island beds or combined with annuals, shrubs and climbers in a way that might shock a purist. Think of Great Comp in Kent. It is certainly a mistake to think that because they come up each year they are trouble-free and need no looking after: they still need weeding and feeding, staking and watering, dividing and replacing.

Some perennials are not wholly herbaceous in retaining some basal leaves in winter rather than disappearing underground altogether: others like Wallflowers or Evening Primroses, though strictly perennial, are usually grown as biennials: others again may not be fully hardy though the ones included in this chapter can generally be relied on to withstand anything in the way of wet or cold that our winter is likely to throw at them, provided they are in suitable soil. They are not confined to the flower garden either: in the herb garden there are Mints, Chives and Lovage; among fruits there are Strawberry and Cape Gooseberry; in the vegetable plot Asparagus and Rhubarb, and we need not distress ourselves by enumerating the perennial nuisances among weeds.

Many genera contain, as well as perennials, annuals, biennials, shrubs and even trees among their species, and weeds as well as desirable plants, so their names may have been dealt with elsewhere. Moreover flowering herbaceous perennials may be included in other chapters as rock plants, ground cover or, most obviously, as bulbs which are both perennial and herbaceous. The seventy or so which follow are quite enough to make up one chapter.

ACANTHUS. The handsome foliage, which is not prickly, is supposed to have inspired the capitals of Corinthian columns and much classical ornament, but the name is composed of Greek *ake* and *anthos*, 'spine-flower' and each one is protected by a spiked calyx. There are several garden species, *A. mollis* being known as Bear's Breeches, which is presumably a misunderstanding of the Tudor herbalist's Branke Ursine from Latin *branca ursina*, 'bear's claws', which the flowers do somewhat resemble.

ACHILLEAS make good cut flowers and contribute cerise and gold clumps to the border. They are named after Achilles, the hero of the *Iliad* who was skilled not only in eloquence and the arts of war but also in healing. Our native white Milfoil *A.millefolium* (thousand-leaf) was highly regarded as a wound herb. Its other name, Yarrow, is of uncertain origin: it was *gearwe* in Old English and there are similar names in German and Dutch.

ACONITUM, Monkshood. There are 300 species and all parts of the plants are poisonous. There are white and blue-and-white varieties as well as the familiar indigo ones. The Greek name used by Theophrastus is of unknown origin but perhaps comes from *akon*, a dart, which was poisoned with Aconite: it is usually applied to the yellow-flowered *A. anthora* (resembling *Ranunculus thora* in its poisonousness) which is not a garden flower. The beauty of the deeply divided leaves should not be overlooked.

ADONIS was a beautiful youth, beloved of Aphrodite, who was killed by a boar he was hunting and, in myth, the red Anemone sprang from his blood. However, the garden Adonises are early-flowering members of the Buttercup family with ferny leaves and glistening yellow bowl-shaped flowers.

AGAPANTHUS, African Lily. The name is composed of Greek *agape*, 'non-sexual love' and *anthos*, 'flower'. Modern hybrids are hardier than the several species from Natal and Cape Province, which were best grown in tubs and moved into shelter for the winter.

ANAPHALIS, Pearly Everlasting. There are 35 species from the Himalayas, Japan and N. America but only four are commonly cultivated. The name is from Greek for an *immortelle* or everlasting flower.

ANCHUSA, Alkanet. These relatives of Borage have the same vivid blue flower-colour. Our native Alkanet was known as *Anchusa sempervirens* (always green) but is now called Pentaglottis (five tongued - look inside the corolla). 'Alkanet' comes via Spain from Arabic *al henna*, 'little henna', because one species in particular, *A. tinctora* from S.E. Europe, was the source of a red dye; it is the name of a cosmetic paint or rouge, though Phillip Miller in his *Gardener's Dictionary*, 1805, says it means 'serpent strangler' because of Pliny's belief that the juice of the plant spat into the mouths of venomous creatures would kill them. Viper's Bugloss, which was credited with the same virtue, is also in the Borage family. At least one modern authority still gives the meaning of Anchusa as 'strangler'.

ARNEBIA, Arabian Primrose. The name is a Latinised form of Arabic *sagaret el arneb*, Prophet's Flower, the story being that the plant broke a fall by Mahomet and the five maroon spots on the yellow flowers are the marks of his fingers.

ARUNCUS, Goat's Beard. The feathery, creamy plumes look splendid at the back of a border, if you have room for it, or by a pool. The name is from the ancient Greek one for the plant and perhaps related to *arundo*, the Latin for a reed.

ASCLEPIAS, Milkweed, Blood Flower. There are 120 species, some shrubby, some tender. If you have a light, neutral soil, *A. tuberosa* is worth trying for its striking orange flower-heads. The name comes from the god of medicine, in Latin Aesculapius, who was worshipped at Epidaurus, and is often depicted with Hygeia, goddess of wise living.

ASTER. To some the name will suggest the annual China Asters, to others the native Sea Aster of our salt marshes or the Michaelmas Daisies which were at first called 'Starworts' when they were introduced to Tudor gardens and 'Aster' means 'star'. Another early name was 'Blue Marigold'. There are 500 species in the genus from all around the world but particularly from N. America, and many splendid garden varieties have been bred from them.

ASTILBE. The name from Greek meaning 'without brilliance' hardly does these beautiful plants justice with their candy-floss plumes and handsome foliage. There is even a variety, *A. taquetti superba*, which is five feet tall. They need plenty of moisture and a little shade to give of their best.

ASTRANTIA, Masterwort. The colours are subtle greenish-white or -pink rather than showy and the form of the leaves, umbels, bracts and star-like (Greek *astron*) flowers are all elegant - a favourite of discriminating gardeners. There is doubt about the origin of the name which may have been coined by l'Ecluse but the English 'Masterwort' comes from the medieval Latin *Magistrantia* and the wort or root which is hot-tasting was used against 'cold' diseases like gout. It has been noticed that it often grows near woollen mills and the seeds may have arrived with imported wool.

CATANANCHE, Cupid's Dart, from the Mediterranean, is like a papery sky-blue Cornflower and is excellent as a cut flower. Its names come from its use by women in love-potions to increase the ardour of their menfolk and the Greek might be translated not too indelicately as 'thrust'.

CAUTLEYA robusta, (formerly called *C.spicata*, spiked) from the Himalayas is exotic looking and not to difficult to grow.It has gold and maroon flower spikes and the whole plant is something like

Sweet Corn.It was named after an army engineer, Sir Percy Thomas Cautley (1802-1877).

CENTAUREA, Cornflower, Knapweed. Our native blue cornflowers are annuals which were hated for blunting the sickles of reapers.Knapweeds get their name from their 'knop' or 'Hardhead'. It is a large genus with about half a dozen perennials useful in the garden.*C. gymnocarpa* (naked fruit) has excellent silver-grey foliage but is only semi-hardy; *C.macrocephala* (big head) is a handsome 4 feet tall with yellow heads and *C. moschata* (musky) which can be white, yellow pink or purple is known as Sweet Sultan. The genus is named after the Centaur Chiron in Greek mythology who cured a wound from a poisoned arrow with one of these plants.

CENTRANTHUS, Valerian, Setwall. *C. ruber*, Red Valerian, has made itself at home here on rocks and old walls. The name from Greek *kentron* means 'spur-flower'. The true Valerians belong to a closely related genus with several species of which *V. officinalis* is adored by cats and a valuable medicinal herb. For the name, see under Corn Salad, (p. 158) .

CHRYSANTHEMUM was the name given by Dioscorides to the Crown Daisy, *C. coronarium*, meaning 'golden flower'. It has become the name of a genus with about 200 species including annuals like our native Corn Marigold, *C. segetum* (of cornfields), and perennials like the Ox-Eye Daisy, *C. leucanthemum* (white-flowered) and Feverfew - probably a native - *C. parthenium*, which got its name, according to Plutarch, by saving the life of a man who fell while building the Parthenon. Tansy, formerly *Tanacetum vulgare*, is now also included in the genus, making four natives. Ox-Eyes were first called 'Marguerites' in 1445 after Henry VI's wife from Anjou who had three of them as her heraldic device. There are garden imports like the Shasta Daisies so called after the place where they were raised in California; a parent was *C. maximum* from the Pyrenees. What most people think of as Chrysanths, however, were bred in China over thousands of years by crossing oriental species, notably *C. sinensis* and *indicus*. The first reached Kew as recently as 1793 and the Pom-Pom and Korean types are still more recent introductions. There are now hundreds of varieties including all the 'incurved' and 'reflexed' monsters of the florist's shop.

CHRYSOGONUM virginianum. Greek *chrysos*, 'gold', refers to the flowers and *gonu*, 'knee', to the jointed stems of this cheerful, long-flowering plant for the front of the border.

COREOPSIS, Tickseed. These are annuals (sometimes listed as Calliopsis) as well as perennials, all from N. America, among these cheerful yellow Daisies. *C. grandiflora* is the most commonly seen. The name means 'bug-like' because the seeds resemble ticks.

DELPHINIUM comes from Greek *delphis*, a dolphin, though the flower bud is actually much more like a tadpole than a dolphin's beaked head. The resemblance is lost in most modern hybrids which have an inconspicuous spur. There are 350 species represented in gardens by the annual Larkspurs and the herbaceous perennials whose parent-in-chief was *D. elatum* (tall), a mountain plant ranging from the Pyrenees to Siberia.

DIANTHUS, Carnation, Gilliflower, Pink, Sweet William. The genus was given its name by Theophrastus and means 'divine flower' or 'Jove's flower' and there is more than one explanation for all the other names. 'Carnation' appears in Turner as 'Incarnacyon', suggesting a flesh-coloured flower or 'God incarnate' but another early spelling is 'Coronation' indicating that they were used for coronets, chaplets or garlands. Gilliflower has been explained as 'July-flower' or even 'Jolly-Flower' but is fairly certainly from French *giroflée* which comes by way of Italian from Latin *caryophyllus* (clove-petalled) and ultimately from the Arabic for a clove. The Clove Carnation is said by Pliny to have been brought from Spain in Augustus Caesar's time and used for flavouring wine. Sops-in-wine is one of the Tudor names.

With Pinks one thing is certain and that is that the name for the colour comes from the plant and not vice versa: what we call 'pink' was 'blush' or 'flesh' or 'rose' or 'light red' up until the 18th century. The flower name may be a shortening of 'pink-eyed', meaning half-closed or screwed-up eyes, or perhaps from the fringed petals looking as if they had been snipped with pinking shears. The Latin name, *D. plumaris*, tends to suggest this and one early writer calls them 'feathered gillofers'. The most famous pink (which is white) is Mrs Sinkins, named after his wife by the master of Slough Workhouse who raised it about 1880. Finally, nobody really knows who Sweet William was, possibly St William of Aquitaine since *D. barbatus* comes from S. Europe but Wallflowers were also known as Sweet Williams in the 16th century.

DICENTRA spectabilis, Bleeding Heart, Dutchman's Breeches, Lyre Flower, Lady's Locket, Lady-in-the-Bath, etc. These numerous names show how popular it became in cottage gardens when it was introduced from China in the 1840s. Linnaeus, who only saw a dried specimen, called it *Dielytra* because he thought it looked like the two wing-cases of a beetle but this was later changed to *Dicentra*, 'two spurs'. *D. exima* (distinguished) from N. America is hardier, more spreading but less attractive.

DICTAMNUS fraxinella, Burning Bush, Bastard Dittany. The Elizabethans seem to have confused this plant with Dittany (*Origanum dictamnus*), a type of Marjoram, the name coming from Dicte, a mountain in Crete. *Fraxinella* means 'like a little Ash', referring to the leaves. It is aromatic and the volatile oils given off can sometimes be ignited on a still summer evening so it shares the name 'Burning Bush' with *Kochia scaparia*.

DORONICUM, Leopard's Bane, provides a splash of yellow in the border early on when not much else is in flower. The specific name *pardalianches* means 'leopard strangler'. Renaissance writers often had great difficulty in identifying the plants referred to by classical authors: some seem to have thought this one was a type of Aconite like Wolf's Bane; others more plausibly identified it with the *Dorongi* of Arab physicians, a useful cordial plant, Henry Lyte saying the root tastes like crayfish.

ECHINACEA, Purple Cone-Flower, comes from the U.S.A. where, I believe, it is known as a Cornflower and one species was used by Indians against snake-bite. They were formerly classed as Rudbeckias. The name is from Greek *echinos*, 'hedgehog' because of the prickly appearance of the cone.

ECHINOPS, Globe Thistle. Most of the species come from the Balkan area and it is not difficult to see how the metallic-blue spherical flower-heads suggested a hedgehog curled up: see the previous entry.

ERIGERON, Fleabane. You might find three Fleabanes growing wild in Britain but one is a rare Alpine and the seeds of another arrived from Canada, it is said, in a stuffed bird and then escaped to spread mightily. The garden varieties, however, are mostly from N. America and difficult to distinguish from Michaelmas Daisies apart from flowering earlier. Theophrastus's name comes from Greek *eri*, 'early' and *geron*, 'old man' either from the fluffy white seeds or perhaps because some species look hoary in Spring.

ERODIUM. There are 90 species ranging from sub-shrubs to rock plants. They are closely related to the true Geraniums or Cranesbills but Greek *erodios* is a heron so these are properly Heronsbills, though sometimes called Storksbills by garden writers who do not know better.

ERYNGIUM, Sea Holly. Apart from the Astrantias, these are the only umbellifers to find a place in the flower- as opposed to the herb-garden. The roots of our native *E. maritimum* were candied and eaten as sweetmeats: Falstaff refers to 'eringoes' as 'kissing comfits' in the last scene of *The Merry Wives of Windsor*. Theophrastus used the name *eryggion* of a spiny-leaved plant, probably *E. campestris*, but its etymology is otherwise unknown.

FILIPENDULA. Our native Meadowsweet, *F. ulmaria* (elm-like) and particularly Dropwort, *F. vulgaris*, are beautiful native members of the genus. *F. rubra* from the U.S.A. and *F. palmata* from Siberia are pink-flowered. The Latin name comes from the way the root-tubers seem to be hanging (*pendulus*) by a thread (*filum*). The Spiraeas are close relatives.

GERANIUM, Cranesbill (Greek *geranos* is a crane) should not be confused with *Pelargonium* (*pelargos* is a stork). Garden varieties may be more compact in habit but are scarcely more beautiful than our native blue Meadow Cranesbill, *G. pratense*, magenta Bloody Cranesbill, *G. sanguineum* or the Dusky Cranesbill, *G. phaeum* of alpine meadows.

GEUM comes from Greek *geyo*, 'to give relish or stimulate the taste' because Herb Bennet, *G. urbanum* (of towns), has an aromatic root with this virtue. Garden varieties mostly derive from

G. chiloense from the island of Chiloe off the coast of Chile, though our native Water Avens, *G. rivale* (of river banks) deserves a place in a damp corner. 'Avens' from medieval Latin *avancia* is of unknown origin.

GYPSOPHILA, Baby's Breath. Border species most often grown are the annual *G. elegans* and the perennial *G. paniculata*. There are also dwarf alpine species. All these relatives of the Stitchworts like well-drained alkaline soil as the name, meaning 'chalk-lover', tells you.

HELENIUM. The Greek name *helenion* after Helen of Troy originally belonged to a quite different plant. Most cultivated varieties are forms of *H. autumnale* from Canada and the U.S.A. where it is sometimes called 'Sneezewort'.

HELIOPSIS meaning 'sun-like' is closely related to the Sunflowers (Helianthus). There are good perennial varieties of *H. scabra* (rough), having leaves which are both saw-edged and bristly to touch.

HEMEROCALLIS. Day Lilies were cultivated in China from antiquity and have been grown in England since Elizabethan times but have recently been much improved in flower size and range of colour. The name is from Greek *hemera*, 'day', and *kallos*, 'beauty', because the individual flowers are short-lived though the plants are hardy and trouble-free.

INULA. Garden Inulas come from Asia and provide yet more golden Daisies for the border. Native species are Elecampane (from Latin *enula campana*) and Ploughman's Spikenard which has a fragrant root. 'Inula', Pliny's name for Elecampane has been explained as a corruption of *Helenula*, 'little Helen' (see *Helenium*).

LIATRIS. The origin of the name of these flowers from the U.S.A. seems to be unknown but colloquially they are known as Blazing Star, Prairie-Button and, most descriptive, Kansas Gay Feather.

LIGULARIA. They used to be classed as *Sencios* and do resemble giant Groundsels which need a damp place. *Ligularia* means 'strap-like' from the flowers' outer or 'ray' petals.

LINARIA, Toadflax. There are useful annuals and alpines as well as the tall purple perennial which seeds itself so readily that it has become naturalised. *Linaria* comes from Greek *linon*, 'flax', (from which linen is made) but so does the genus *Linum* to which true Flax belongs, so Toadflax (*Krottenflachs* in German) is 'useless flax'.

LIRIOPE muscari. As the specific name suggests, this plant somewhat resembles a large Grape Hyacinth but the flower-spikes appear in late summer. The name is that of the woodland nymph in Greek mythology who was mother of Narcissus.

LUPINUS. At first annual European species were grown in English gardens like the yellow-flowered *L. luteus* which was called 'Spanish Violet' because of its scent. Perennial species first arrived from Virginia in the 17th century, Tree Lupins from California in the late 18th, but it was not until 1826 that *L. polyphyllus*, one of David Douglas's discoveries, arrived from British Columbia. This was the parent-in-chief of modern garden Lupins though their true potential was only developed by George Russell in the present century. The Romans grew *L. alba* as a fodder crop and called it *pisum lupinum*, 'Wolf Pea'. The seeds, though bitter, can be used as human food if boiled in several changes of water and it has been suggested that the name may originally have come from Greek *lupe*, 'grief', because they made one grimace.

LYCHNIS. Campion, Lamp-Flower. Greek *lychnos* is a lamp and it has been suggested that the flowers were thought to be bright as a lamp flame. I doubt this explanation. *L. coronaria* seems to have been the first to be given the name and its flowers are magenta, not flame-coloured. It has woolly leaves which might have been used as lamp-wicks. The specific name *coronaria*, 'making a crown', may explain the name Campion: Champion and Campion were interchangeable spellings so perhaps the flowers were used for crowning champions. Another useful perennial, the scarlet *L. chalcedonica* is a native of Russia but was probably introduced to Turkish gardens (Chalcedonia is the Bosphorus area) and brought to Western Europe by Crusaders. It became popular and some of its popular names strongly suggest this history, e.g. Flower of Constantinople, Jerusalem- or Maltese-or Knight's-Cross. It is also known as Bridget-in-her-Bravery, Nonesuch and Flower of Bristow, the last because its colour resembles a dye known as Bristol Red.

LYTHRUM. There are several good, easy perennials which can be pink or red as well as our native Purple Loosestrife, *L. salicaria*, (willow-like). The name is from Greek *lythron* meaning black or arterial blood. For 'Loosestrife' (see p.19) .

MACLEAYA, Plume Poppy, Tree Celandine. The handsome foliage and plumes of cream or pinkish flowers rising to six or eight feet make this a subject for the back of a large border. Alexander Macleay (1767-1848) was Colonial Secretary for N.S. Wales and Secretary of the Linnaean Society in London.

MECONOPSIS. These Poppies are hardy but tend to be short-lived. This does not matter with the Welsh Poppy, *M. cambrica*, as it flowers and seeds itself so freely, and the tall blue Poppies from the Himalayas, *M. betonicifolia* (Betony-leaved) and *M. grandis* are so splendid that they are worth any amount of trouble. *Mekon* is 'Poppy' in Greek and *-opsis* means 'resembling'.

OPHIOPOGON with its grass-like, almost black leaves is intriguing rather than beautiful and its spikes of lilac-pink lily flowers are not showy. The name means 'serpent's beard' but the reason it acquired it is obscure.

PENSTEMON means 'five-stamens' and one of the many species of this genus from N. America, *P. barbatus*, gave rise to the English name 'Beard Tongue'. Another species, *P. digitaloides* reminds us that there is something Foxglove-like about the flower-spike of many of them, even the dwarf alpine ones.

PHLOX is a name used by Theophrastus for some 'flame-coloured' flower but he could never have seen our Phloxes which come from N. America. Border Phloxes, mostly bred from *P. paniculata*, are invaluable for late summer scent and colour which is seldom really flame-like. There are, of course, lovely Rock Phloxes and the *drummondii* annuals which were collected in Texas by Thomas Drummond about 1835.

PHYSOSTEGIA is sometimes called Obedient Plant because you can move around the pink or white tubular flowers on their tall spikes without them breaking off or springing back. The Greek name, however, comes from the inflated calyx and means 'bladder-covering'.

PLATYCODON, Balloon Flower, Japanese Bellflower. The open flower looks like a small five-petalled Clematis but it is presumably the fascinating balloon-like flower buds (or possibly the corolla) that gives the name from Greek *platys*, 'broad' and *kodon*, 'bell'.

POLYGONATUM, Solomon's Seal. As with the Knotweeds, the Greek name means 'many joints'. It was once reputed to cure rheumatics and bruises. The English name is more problematical: the flowers do resemble seals dangling from a chain or document but Solomon's seal should be the five-pointed Star of David or pentangle. Prominent leaf-scars on the stems might be thought to resemble seals but not, I think, stars.

POLEMONIUM caeruleum (sky-blue), Jacob's Ladder, Greek Valerian, is native in the north of England though now uncommon. I have only seen it growing around Thor's Cavern in the Manifold Valley in Staffordshire. It has other names like Blue Jacket, Poverty, Charity, and Ladder of Heaven. There are better species for the garden like *P. foliosissimum*, (very leafy), from the Rocky Mountains. Pliny uses the name *Polemonium* for a medicinal herb: *polemos* means 'war' and he says two princes came to blows about who discovered the plant. It is also said to be named after King Polemon of Cappadocia.

POTERIUM is the name of our native Salad Burnet and in Latin means a drinking cup. *P. obtusa*, (blunt or rounded, referring to leaflets), from Japan has a pink flower spike with long stamens which make it look like a bottle-brush. It is sometimes known (like the Bloody Burnet) as *Sanguisorba*, (blood-stauncher).

PYRETHRUM roseum was introduced from the Caucasus early last century and today's red, pink and white single and double Daisies with carroty foliage have been bred from it. Botanically it is *Chrysanthemum coccineum* (scarlet). Long ago the Persians used it pulverised as an insect powder and today it is an important crop in parts of Africa for use in insecticides. The name comes from Greek *pyr*, 'fire' because several Chrysanthemum species have a hot-tasting root which was used against fevers: Feverfew is one of them.

RODGERSIA. Several species are grown as much for their handsome Horse Chestnut-like foliage as for their flower plumes. They need moisture and a bit of shelter. Rear Admiral John Rodgers was an American naval officer who commanded the Pacific Expedition, 1852-6, when the genus was discovered.

SIDALCEA. False Mallow. There are 25 species but *S. malviflora*, (with Mallow-like flowers), has provided most of the pretty garden varieties. The name is composed from two now defunct genera, *Sida*, the Greek name for a water lily, and *Alkaia*, a kind of Mallow.

SOLIDAGO, Golden Rod. The specific name of our native *S. virgaurea* means precisely 'rod of gold' and *Solidago* is from the Latin verb *solidare*, to unite or make whole, from its use as a sovereign wound-herb and for 'inward hurts' including 'gravel, stone and stranguary'. Cultivated varieties are mostly from N. American species and have become much more compact and less weedy of late though the colour does not stray far from golden yellow.

THALICTRUM, Meadow Rue. We have several native species and several more have been introduced to gardens, the best being *T. dipterocarpum* (two-winged fruit) which was found by the Abbé Delavay in W. China. They all have delicate blue-green foliage and fluffy flowers on wiry stems giving them something of the character of the Gypsophilas. The name was used by Dioscorides from Greek *thallo*, to grow green or flourish.

TROLLIUS, Globe Flower. These relatives of our Buttercups and Marsh Marigold need a moist soil and *T. europaeus* is native in N. Britain and known in Scotland as Lucky Gollans. Garden varieties are hybrids with Asian and Far Eastern species. The German name is *Trollblume* from *Trol*, a globe, and this is usually said to be the origin of *Trollius*, but Latin *trulleus* is a basin and *trulla* a ladle for transferring wine from bowl to cup.

VERBASCUM, Mullein. Our native Mulleins are biennials and have a host of country names like Adam's Flannel and Beggar's Blanket from the felted leaves of the common species. There are striking perennials for the border as well and colours include white, plum purple as well as yellow and the unusual biscuit shade of *V. hartleyii*. 'Mullein' comes via Old French *moleine* from Latin *mollis*, 'soft', and Verbascul, Pliny's name, is said to be a corruption of *barbasculum*, 'bearded', but whether from the woolly leaves, the prominent stamens or for some other reason I do not know.

YUCCAS come from Central America and the sharp-pointed leaves have given rise to popular names like Adam's Needle and Spanish Bayonet. 'Yucca' is the Carib name for the food plant Cassava (*Manihot esculenta*), a source of tapioca, and was mistakenly thought to refer to *Y. gloriosa*. *Y. filamentosa* (having threads - on the margins of the leaves) is most often grown here, as it is the most likely to flower reliably in our climate.

Chapter 7
A Pick Of Pot Plants

Almost anything can be grown in a pot. Hardy natives like Daffodils and Primroses to brighten the darkest days of winter are indispensable and so is Mignonette for its scent in summer. This chapter is not about things like that. My bathroom is currently being taken over by a Persimmon tree grown from a stone and known as Percy, and he has been preceded by Avocado, Date Palm, Carob and other exotica. Some people keep unlikely pets and some gardeners cannot resist having a go with any seed that comes their way. Perhaps a book should be written specially for them, but this is not it.

Very many tropical plants are grown by nurserymen into spectacular pot plants, but from the moment they leave the controlled conditions of the grower's greenhouse for the florist's shop and the living room, their days are numbered. Draughts, dry air or inadequate daylight will do for them sooner or later even if we keep them warm enough and remember to repot, and to feed and water them regularly, not too much and not too little. At least they last longer than cut flowers and the most satisfactory are the ones that take longest to die and set us an example by looking cheerful while they are about it. Plants from the floor of the tropical forest often adapt well to our buildings because they have learnt to put up with dim light and warm temperatures. Humidity is usually the missing ingredient necessary for their comfort and a moist gravel tray to stand them in can make all the difference between failure and success. Other things are adapted to semi-desert conditions and will hate wet feet and poor light, but thrive in central heating and shrug off dust: it always pays to know where your plant comes from.

Clearly this chapter should really have been called 'A Long Suffering of House Plants'. To keep it to manageable proportions I have had to leave out scores of lovely things like the *Odontoglossum* (tooth-tongued) Orchids, the Amazon Lily, *Eucharis* (agreeable) or the *Fittonias* (named after Eliza and Sarah Fitton, who wrote '*Conversations about Botany*' in 1817) because they all really need a warm greenhouse to be at all happy. For the same reason, only a very few of the best known Cacti, Orchids and Ferns are included. Popular pot plants which have close relatives hardy enough to be grown out of doors are likely to have had their names dealt with elsewhere; others again are in the 'Buddle & Fuchs' and 'Bulb' chapters.

ACALYPHA is the classical name for the nettle and it was given to this genus by Linnaeus because of a supposed similarity of the leaves. *A. wilkesiana* is known as Copperleaf or Beefsteak Plant and was named after the American Admiral-explorer of the Pacific, Charles Wilkes (1798-1877). *A. hispida* (bristly) is known as Chenille Plant or Red Hot Cat's-Tail from its drooping tassel-like flower spike. Both need warmth, humidity and good light.

ADIANTUM, Maidenhair Fern. The name comes from Greek *adiantos*, 'unwetted' because of the remarkable way the leaves repel water. There are 200 species, the best known being *A. capillus-veneris*, a rare native in the milder S.W. parts of Britain. The name can be translated as 'Venus' hair' from the delicate foliage or, more probably, 'pubic hair' from the wiry black stems.

AECHMEA, see under Bromeliads.

AGLAONEMA is from the Greek meaning 'bright thread' referring to the stamens. *A. modesta* (modest) is popular because easy to grow and known as Chinese Evergreen but there are better species available like *A. commutatum* (changeable) with silver-green variegation. They are all shade lovers.

ALLAMANDA, Golden Trumpet, needs warmth as it comes from tropical S. America. It is named after the 18th century Swiss botanist who sent seeds of *A. cathartica* (purgative) from Surinam to Linnaeus round about 1770. *A. violacea* is violet-flowered but less free flowering.

ALOE. The name comes from Latin and Greek for the plant and its bitter juice which has a long history of medicinal use. There are many species of which the most attractive is *a. variegata* from S. Africa known as the Partridge-Breasted Aloe.

ANANAS, see Bromeliad.

ANTHURIUMS come from tropical S. America. The name

from Greek means 'flower-tail' referring to the spadix which is usually white or pink in a scarlet spathe. There are many species, the two most commonly grown being *A. scherzerianum*, known as Flamingo- or Flame-Flower, named after its discoverer, a Viennese called Scherzer, and *A. andreanum*, the Pigtail Plant or Painter's Palette, named after E.F. André (1840-1911).

APHELANDRA squarrosa from Brazil is known as Saffron Spike or, more commonly, as Zebra Plant from the handsome striped leaves. The generic name means 'simple male' from the one-celled anthers and *squarrosa* seems to mean something like 'rough, sticking out in all directions', referring to the cone-like spike of yellow bracts between which the flowers appear.

ASPIDISTRA. The name comes from Greek *aspideon*, a buckler, because of the mushroom-like shape of the stigma, and, since they are almost impossible to kill by neglect or poor light, they are sometimes known as Cast Iron Plants. *A. elatior* (taller) from Japan is the only species grown and there is a slightly less robust variegated form with a broad cream band on the leaf.

ASPLENIUM, Spleenwort, is a huge genus of evergreen ferns which takes its name from the ancient use of some of the species against diseases of the spleen and, according to Culpeper, "strangulary, stone in the bladder, jaundice and hiccough", but it couldn't be given to women because it caused barrenness. We have ten native species of which the Maidenhair Spleenwort with its green zip-fasteners growing on limestone rocks or in old mortar is probably best known. This is *A. trichomanes*, Theophrastus' name meaning 'hair-madness'. One of the most popular species as a pot plant is *A. nidus*, the Bird's Nest Fern from tropical Asia, but several others are also available.

BELOPERONE guttata (spotted, dotted with glands), Shrimp Plant. The name comes from Greek *belos*, 'arrow', and *perone*, 'fastening', because of the shape of the junction between stamen and anther lobes. It is now correctly called *Justicia bandegeana* after the 18th century Scottish horticulturalist, James Justice, and the American civil engineer who studied plants in Mexico, Townsend Stith Bandegee (1843-1925). It is the red or yellowish bracts borne at the tips of the stems that are somewhat shrimp-like.

BILLBERGIA. The genus contains some of the easiest and most popular *Bromeliads* (see below) and was named after J. G. Billberg (1772-1844). *B. nutans* (nodding) is known as Queen's Tears and *B. horrida* (spiny, the leaves) is a large plant with a more upright flower spike. In both, blue-tipped flowers contrast strikingly with pink bracts. Today you are very likely to find modern hybrids offered for sale.

BROMELIADS. The big family *Bromeliaceae* comes mainly from C. and S. America and contains epiphytes from the tropical rain forest, but also desert and sea-shore plants. It was named after the Swedish botanist, Olof Bromel (1629-1705). Many form water-holding cups so there are so-called 'Urn Plants' in several of the genera.

ACHMEA from Greek *aichme*, a point, from the sepals: *A. fasciata* (banded) is one of the 'Urn Plants'.

ANANAS is the native Brazilian name for the Pineapple and *comosus* is 'cultivated': '*Variegatus*' makes the best pot plant but you'll need a greenhouse to get fruit.

BILLBERGIA (See above).

CRYPTANTHUS (hidden flower), Earth Star. Many of the best species have longitudinally-striped leaves.

GUZMANIA is named after the 18th century Spanish naturalist Anastasio Guzman. They are mainly rain forest epiphytes and need humidity.

NEOREGELIA is named after E.A. von Regel (1815-1892). The genus contains the 'Blushing Bromeliads' whose leaves change colour as flowering approaches, and also *N. marmorata*, the Marble Plant.

NIDULARIUM from Latin *nidus*, a nest, because of the cup in which the bracts and small flowers appear.

VRIESIA after the Dutch botanist W.H.. de Vriese (1806-1862). Many species have leaves with transverse bands of darker green or purple, but also bright coloured flower spikes.

CALADIUMS, known as Angel's Wings or Heart of Jesus, come from the tropical forests of S. America. The heavily veined pink, cream, red and green leaves will die down to the tuberous root for a resting period each year. The name is a Latinised form of the native one, *Kaladi*. Usually grown today are hybrids of *C. bicolor* and *C. hortulanum* (of gardens).

CATTLEYA is a genus of some 60 epiphytic orchids from C. and S. America including the so-called Christmas and Easter Orchids. It is named after William Cattley of Barnet who was a patron of horticulture and collector of rare plants in the early 19th century. The many hybrids contain some of the easiest Orchids for the amateur and the most familiar at the florist's.

CERATOSTIGMA, see Plumbago.

CHAMAEDOREA, see Palm.

CHAMAEROPS, see Palm.

CHLOROPHYTUM is Greek for 'green-plant'. The small plantlets which weigh down the arching stems give this well-known house plant its English name of Spider Plant and also the most commonly grown one with its specific name *comosum* (tufted). The variegated form is nearly always seen and it needs good light or the leaves will brown. Also grown is *C. capense* (from the Cape) with larger starry white flowers.

CISSUS rhombifolia (diamond-shaped leaves), Grape Ivy. The name is from Greek *kissos*, 'ivy', but the genus belongs to the Vine family. Toughest and least complaining is the dark green, glossy *C. antarctica*, actually from E. Australia and known as Kangaroo Vine. A much handsomer climber is *C. discolor* (two-coloured) with leaves marbled like *Begonia rex*, and other species are available. I have never seen flowers on any of them.

CITRUS was the Latin name of the Citron but Linnaeus gave it to the whole genus. It is easy enough to grow the pips of Oranges and Lemons in pots but, if you want flowers and fruit, look for *C. mitis* (bland, not acid), also known as *C. microcarpa* (tiny-fruited), which only grows about 18 inches high.

CLERODENDRON. In Greek the name means 'chance-tree' apparently because of the variable medicinal properties of the species which include many tender trees, shrubs and vines. *C. thomsoniae* was named after the wife of the Rev. W.C. Thomson who worked in W. Africa in the mid 19th century, and is a vigorous climber which needs warmth, but can be controlled by pinching back. It is known as Glory Bower or Tropical Bleeding Heart, the clusters of pendulous flowers having a crimson corolla which emerges from a heart-shaped white calyx.

CODIAEUM variegatum, Croton, is a handsome foliage plant from Malaysia related to the Euphorbias. The name is Latinised from the native one and the variety '*pictum*' (painted) is the one usually seen.

COLEUS. The name 'Flame Nettle' is descriptive but does not seem to be widely known for these popular but short-lived foliage plants. In Greek *koleos* is a sheath referring to the way the stamens are united. The genus is only doubtfully distinct from *Plectranthus* (Spur-flower). *C. blumei*, named after C.L. von Blume (1796-1862), in its many varieties is nearly always grown, though *C. thyrsoides* (staff-like) deserves to be better known: its spike of beautiful blue flowers rises from plain green leaves.

CORDYLINE comes from Greek *kordyle*, a club, from the shape of the fleshy root of some species. *C. terminalis* (from the flower at the end of the stem) is known as the Good Luck Plant or Hawaiian Ti in which country it is used in cooking, to thatch houses and make hula skirts. There are bi- and tri-coloured varieties and it is sometimes listed as a *Dracaena*. *C. australis*, the Cabbage Palm, will grow up to 10 feet in the wild, but can be kept to 2 feet in a pot. *C. indivisa* (having a single stem) is about twice the size.

CRASSULA is a large genus of mainly tender succulents and the name from Latin *crassus*, 'thick', refers to the fleshy leaves and not to any supposed stupidity. Indeed, several are grown as exceptionally tolerant, slow-growing house plants. Best known as the Jade Plant, (*C. argentea* = silver, also called *C. portulaca* = milk-carrier) and *C. arborescens* (tree-like).

CYMBIDIUM comes from Greek *kymbe*, a boat, from the typical hollow lip of this Far Eastern genus of Orchids. Many of the 60 species are rather large, but numerous hybrids, including miniatures have been bred and they are not difficult for the amateur to grow.

CYPERUS. The name comes from the Greek for Sedge and Papyrus belongs to the genus. They are happiest with their feet in water. Most often grown as a pot plant is the graceful Umbrella Grass from Madagascar, *C. alternifolius* (alternate-leaved), and the variegated form of the dwarf variety from Australia, *C.a. gracilis*(slender, graceful) is the most desirable.

CYPRIDPEDIUM, Lady's Slipper, Moccasin Flower. Cypria (*Kypris* in Greek) was one of the names given to Venus (Aphrodite), goddess of love, because the island of Cyprus was one of the chief seats of her worship. The name of this genus of hardy Orchids therefore means Venus'-slipper (*pedilon*). *C. calceolus* (like a small shoe) is one of the rarest of our native species.

DATURA, Angel's Trumpet, Horn of Plenty. There are annual and shrubby species, the latter mostly needing a large conservatory. The name may be from Arabic *tatorali* or from an Indian vernacular name. I have only grown *D. meteloides* (like *metel*, the Chinese name for another species), a short-lived shrubby species from Mexico. Its grey hairy leaves are disagreeably pungent, warning that it is poisonous, but the large white flowers have an elusive but hauntingly sweet perfume. The Thorn Apple (*D. stramonium* = spiky-fruit, Theophrastus' name) pops up from time to time in this country on waste ground but is not a true native.

DENDROBIUM is one of the largest and most varied genera of Orchids with nearly 1000 species. They are mostly epiphytes, hence the name meaning 'tree-living'. Many of the tropical species need the humidity and warmth of a heated greenhouse but there are others, like *D. nobile* (notable) from the Himalayas, which can be happy on a living room window sill.

DIEFFENBACHIA is known as Dumb Cane or Mother-in-law's Tongue because the leaves contain crystals of a poison, calcium oxalate, which can cause temporary speechlessness if they are chewed. There are many species, hybrids and cultivars, some becoming rather large. Varieties of *D. maculata* (with white-blotched leaves) are among the smaller and most popular ones. They come from tropical America and were named after J.F. Dieffenbach (1790-1863) who was in charge of the gardens at the Schönbrun Palace in Vienna.

DIZYGOTHECAS are known as False Aralias because they

were once included in that genus which has the French-Canadian name *aralie*. *D. elegantissima*, the Finger Aralia from the New Hebrides, has palmate leaves with long strap-like saw-edged segments which are copper-red at first before turning dark green. The name is composed of *dis* (twice), *zygos* (a yoke) and *theka* (case) and refers to the fact that the anthers have twice the usual number of lobes.

DRACAENAS make tough Palm-like house plants which can become large. Indeed, a specimen of *D. draco*, the Dragon Tree, in the Canary Isles, was 70 feet tall and reckoned to be 6000 years old when it was blown down by a gale in 1868. Several other species are grown, the commonest and easiest being *D. marginata*, which has red margins to the leaves when the light is adequate. The genus probably gets its name from Greek *drakaina*, a she-dragon, because the milky sap when dried becomes a reddish gum like the resinous substance known as 'dragon's blood' which is obtained from a Malaysian Palm and has medicinal and other uses. Another theory is that Clusius, who lived in Flanders, may have named it in 1581 in celebration of Drake's victory over the Armada.

EPIPHYLLUM, Orchid Cactus. It is a mistake to think of all Cacti as desert plants. This genus (like the Christmas Cacti, *Schlumbergera*) come from tropical America and have no need of prickles, as they grow epiphytically in the canopy of the rain forest. Modern hybrids provide a range of colour and the large flowers are borne upon the notched, flattened stems. These were once mistaken for leaves, hence the misnomers 'Leaf-Cactus' and *epi phylon* (=upon the leaf).

EXACUM affine, Arabian or Persian Violet, is a pretty little biennial from the island of Socotra at the mouth of the Gulf of Aden, whose fragrant white or violet flowers are a bit like a Potato's. The name apparently comes from the Latin *ex ago*, 'I drive out' but was transferred to this genus in the Gentian family from a *Centaurium* which was supposed an antidote to poisons. The specific name *affine* means 'related to' (affianced).

FATSIA japonica is *Fatsi* in Japanese. It makes a trouble-free house plant and will survive out doors in the mildest areas. The same can be said for *FATSHEDERA lizei*, a chance bigeneric hybrid that was spotted in the Lizé brothers' greenhouse in France just before WWI. Its parents were *F. japonica 'Moserii'* and the Irish Ivy, *Hedera helix 'Hibernica'*. The leaves are Ivy-shaped rather than Fig-like and there is a variegated form.

GYNURAS from Java have glossy young leaves which are wine-red underneath and covered on top with velvety purple hairs. *G. sarmentosa* (having long stolons or runners) is known as the Purple Passion Vine and quickly becomes straggly if it is not frequently pinched back. The Groundsel-like flowers are not very pleasantly scented. *G. aurantiaca* (orange-coloured, the flowers), the Java Velvet Plant is similar.

KALANCHOE. There are many tropical species of these succulents, the name being the Chinese one for them. Nearly always sold today at Christmas time are red, orange or yellow-flowered cultivars derived from *K. bossfeldiana* from Madagascar, named after Robert Bossfeld, a German nurseryman.

LACHENALIAS are easily grown bulbs from S. Africa named after Werner de la Chenal (1736-1800) who was Professor of Botany at Basle. The usual English name is Cape Cowslip but they are more like little loose-flowered Red Hot Pokers, Muscari or Aloes as the name *L. aloides* suggests. Several other species are grown such as *L. bulbifera* (having bulbs), *L. glaucina* (having a waxy bloom) and *L. mutabilis* (changeable - the colour of the flowers).

MARANTAS are named after a 16th century Venetian botanist, Bartolomeo Maranti. Arrowroot is a member of the genus but the species grown for its beautifully marked leaves with a prominent fish-bone pattern of veins is *M. leuconeura* (white-veined) from Brazil. It is called the Prayer Plant because these leaves are horizontal in the light, but after dark stand upright as if folded reverently: '*erythroneura*' is a red-veined variety.

MONSTERA deliciosa, Swiss Cheese Plant, Shingle Plant. The name is of uncertain origin but probably connected with Latin *monstruosus*, 'strange, wonderful, monstrous', in reference to the odd incisions and perforations developed by the mature leaves which, of course, suggest Gruyére cheese. It is sometimes wrongly called the 'Split-leaved Philodendron'. The genus contains 50 tropical climbers. In the wild *M. deliciosa* produces a club-shaped fruit which is said to taste deliciously of Banana and Pineapple. In Mexico, where it comes from, plant and fruit are called Ceriman.

NIDULARIUM, see Bromeliads.

PALMS belong to several plant families and are mostly too big for greenhouses, let alone pot culture, but there are three exceptions. *Chamaedora elegans* from Mexico and Guatamala is the once-popular Parlour Palm or Dwarf Mountain Palm which seems to be coming back into favour. The generic name comes from Greek *chamai*, 'on the ground' and *dorea*, ' gift', because the fruit is borne low down within reach. *Chamaerops humilis*, the Dwarf Fan Palm, is the only Palm native to Europe. Greek *chamai* and Latin *humilis* could both be translated as 'low growing' and *rhops* is a bush. *Phoenix roebelinii*, the Miniature Date Palm, was collected in S. E. Asia by M. Robelin. The generic name probably comes from the fact that the Greeks had the Date Palm from the Phoenicians and is nothing to do with the fabulous bird of Egypt.

PEPEROMIA. The 1000 species come from the W. Indies and S. America. The name means 'like Pepper' and they are closely related to the true Pepper, *Piper nigrum*, of the E. Indies. Several species are grown as bottle garden or pot plants: most familiar are *P. caperata* (wrinkled) with corrugated, heart-shaped leaves and white mouse-tails of tiny flowers borne above them on pink stalks and also *P. argyreia* (silvery), known as the Water Melon or Rugby Football Peperomia from the shape and markings on the smooth leaves.

PHALAENOPSIS from Greek *phalaina*, a moth, is well named. There are about 35 species of Moth Orchid and modern hybrids are among the most beautiful and least difficult for the non-specialist to grow.

PHILODENDRONS love (*phileo*) to climb a tree (*dendron*) in Brazil and Columbia, at least, many species in this big genus do. They are grown as pot plants for their attractive foliage but will seldom flower in captivity. One of the best known is the Heart-Leaved Philodendron, *P. scandens* (climbing) which is also called *cordatum* (heart-shaped) and *oxycardium* (acid-heart). This was first introduced into England from Jamaica by Captain Bligh of The Bounty. *P. erubescens* (blushing) has purple leaf-stems and coppery leaves and there are bipinnate species known as Saddle Philodendrons or Panda Plants.

PHOENIX, see Palms.

PILEAS are spreading members of the Nettle family, the best known having herringbone-patterned leaves. The name probably comes from the Latin *pilleus*, a Roman felt cap, and refers to the calyx covering the seed. The Aluminium Plant, *P. cadierei* was collected in Thailand by R.P. Cadiere as recently as 1938. The dwarf variety 'Nana' stays compact. *P. involucrata* is similar but has coppery leaves with red undersides and the name indicates that the leaves cluster beneath the flower like an involucre or ring of bracts. I have not discovered why it is sometimes known as the Friendship Plant. *P. microphyla* (with tiny leaves) is quite different and gets its name of Gunpowder or Artillery Plant because the anthers emit tiny puffs of pollen from the flowers if the plant is touched.

PLATYCERIUM means 'broad-horn' and they are impressive epiphytic ferns which generally want the humidity of a tropical forest. *P. bifurcatum* (forked) from N. Australia is the least fussy and is well named the Elk's-horn or Stag's-horn Fern.

POINSETTIA. The well-known Christmas pot plant with its scarlet bracts is now properly called *Euphorbia pulcherrima* (most beautiful). Joel R. Poinsette (1775-1851) was a diplomat as well as a gardener and botanist, and it was when he was American Ambassador to Mexico that he sent this plant to friends in S. Carolina.

POLIANTHES tuberosa from Mexico gets its name 'Tuberose' because it is tuberous and not because its pearly white flowers smell or look like a rose (though they are heavily scented). *Polianthes* means 'grey-flower' (Greek *polio*) presumably describing other species in the genus which are not cultivated: only an ineluctable depressive would want a pot of grey flowers.

RHIPSALIDOPSIS. The Easter Cactus is *R. gaertneri* after K.F. von Gaertner (1772-1850), a pioneer of hybridization, and is often confused with the Christmas Cactus, *Schlumbergera*, but has bright red flowers in Spring. The genus gets its name from the supposed resemblance of the interlacing twigs to wickerwork which is *rhipsalis* in Greek. There is one other species which has pink flowers a month or two later.

RHOICISSUS: another name for Grape Ivy, see Cissus.

SCINDAPSUS is the Greek name for an Ivy-like plant and *S. aureus* (with gold-splashed leaves) from the Solomon Isles is known as Devil's Ivy. It is also known as an *Epipremnum* (upon a tree stump) and *Rhaphidophora* (needle-bearing) and makes a tolerant house plant, a bit vigorous if allowed to climb but more manageable trailing. *S. pictus* (painted), Silver Vine from Borneo, has silver-edged leaves. There are named varieties of both and they tend to lose their variegations in poor light.

SPATHIPHYLLUM wallisii is known as the Peace Lily or White Sails and comes from Columbia whence it was introduced in 1824 by Gustave Wallis. The name means 'leaf-like spathe', the flowers being like a Cala Lily or its even closer relative *Anthurium* which also comes from Columbia.

STREPTOCARPUS. If you have ever let them go to seed, you will have seen the twisted (*streptos*) seed pod (*karpos*) which is like a corkscrew. The name Cape Primrose refers to their S. African origin and somewhat Primrose-like puckered leaves. The blue or mauve flowers of some species are actually more like large Violets than those of their relatives the African Violets, *Saintpaulia*, while other species remind one more of *Gloxinias*, another relative in the *Gesneria* family. Generally grown today are hybrids, the 'Nymph' varieties with various girls' names having the virtue of flowering for many months.

ZEBRINA pendula (drooping) from Mexico is often called *Tradescantia* because it resembles the tender, trailing species in that related genus and both are known as Wandering Jew. *Zebrina* is the more colourful, the leaves being purple beneath and having two silver stripes (Latin *zebra*) above.

Famous herbalists
Those in front are thought to be: l to r. Pliny, Dioscorides and Serapion. Frontispiece of Peter Schöffer's *Der Gart*, 1485

Chapter 8
A Sheaf Of Shrubs

'Shrub' is not an attractive word: in origin it was the same word as scrub and its first meaning was rough or coarse. It came to mean the undergrowth of a wood and suggested stunted and prickly boscage. To call somebody 'scrubby' would not be complimentary, yet it was once the name of an important chieftain. Shrewsbury and Shropshire were originally Scrub's fort and shire. He may have been an Anglo-Saxon or perhaps a Norman with an English nickname. Anyway, he was the father of the Richard who built Richard's Castle in Herefordshire. A collection of consonants like 'scr' at the beginning of a word looked impossibly barbarous to Norman scribes who set about civilizing the spelling.Provided they are treated with respect and disciplined, there need be nothing scrubby or uncivilized about shrubs. Indeed, they are an indispensable part of the glory of the garden providing shelter and screening, colour and scent with delights of flower and foliage, bark and berry. Think of Roses and Rhododendrons, Lilacs and Lavender, Heathers and Hydrangeas.

In the eighteenth and early nineteenth centuries the 'Wilderness' or shrubberies were of native evergreens, Box, Yew and Holly to which Common and Portugal Laurels, Acubas and Skimmias were later added. They furnished the edges of the garden, providing shelter and privacy and could be threaded by paths for gentle exercise protected from excesses of wind or sunshine. They made the ideal place for an assignation, an elopement or even a murder. Later still came the Rhododendrons and all manner of flowering things from every corner of the globe. Many a great garden and park was jazzed up in a way that would have struck its designer - Brown or Repton perhaps - as shockingly lacking in dignity. The smaller garden did not have room for the traditional shrubbery and the wealth of newly available plant material was a temptation to make overcrowded shrub-borders into rag-bags of bushy things. No class of plants gets worse treated, jammed together without room to breathe or hastily stuck into draughty corners and subsequently neglected.

It is not too difficult to define a shrub: 'a woody plant smaller than a tree usually with several major branches arising from or near the base of the main stem.' It may be less easy to decide in marginal cases: what about scramblers like Clematis, squatters like Rock-Rose, or demi-trees like some of the Sorbuses and Magnolias? Some trees can and often should be kept shrub-sized by regular pruning: *Eucalyptus gunnii* would be an example; the golden Robinia another. Some can be made into a hedge or a maze, potted or pleached, persuaded to climb a trellis or form an arch, clipped into a pyramid or a peacock or subjected to worse indignities. We will leave the scramblers and climbers and the trees for other chapters. A few will take their proper place among the herbs and rock-plants, and a large number named after the people who discovered them, introduced them to cultivation or were otherwise honoured in their naming will be found in Chapter 4. Before we proceed to the remainder, however, a word had better be said about some disgraceful goings-on in Ancient Greece.

A lot of trouble could have been avoided on Mount Olympus if Paean, physician to the Gods who gives his name to our Paeonies, had prescribed stiff doses of sedatives all round. The nymph Daphne had to turn herself into a shrub to avoid the unwelcome attentions of Apollo so that Daphne came to be the Greek name for what the Romans were to call *Laurus nobilis*, the Bay Tree or Noble Laurel. Our native Spurge Laurel (*Daphne laureola*) appears to have a tautological name, 'the Laurel-like Laurel', but Bay leaves were used to make wreaths for the brows of victors so it translates more sensibly as 'the Laurel of garlands'. It has evergreen leaves which were thought to resemble Bay but it is quite unrelated. It belongs to the family *Thymelaeceae* or 'Thyme-Olive' like our other rare native, the lovely, fragrant, pink-flowered *Daphne mezereon*, which doesn't even have evergreen leaves to make it Laurel-like. Its specific name is the Arabic one used by the great physician Avicenna whose writings were the standard medicinal reference work throughout the Middle Ages. The evergreen shrubs that most people call Laurels are also unrelated to the Bay-Tree and, therefore, not true Laurels either. The Common or Cherry-Laurel (*laurocerasus*) and the Portugal

Laurel (*lusitanica*) are both in fact evergreen Plums (*Prunus*): that comes as a surprise to many gardeners, I find.

Syringa commemorates another near rape: the goat-footed God Pan (who inspired panic among the girls) pursued the nymph Syrinx into a river. To make doubly sure that he was properly cooled off, she turned herself into a reed. All he could do in his frustration was to pluck it and turn it into Pan-pipes. The Mock-Orange, which was known to the old herbalists as the White Pipe-Tree came to be called Syringa because of its hollow, pithy, reed-like stems which make even better blow-pipes than Elder. Nowadays, however, we have to reserve the name 'Syringa' for the Lilacs, though they do not have reed-like stems, and call the Mock Oranges by the Greek name *Philadelphus* meaning 'brotherly love', something that was in short supply in the bloody old myths.

With those confusions resolved and with mythical concupiscence safely behind us we can now proceed to less exciting subjects.

ABUTILON is from the Arabic name for a similar plant. The lampshade-like flowers are a clue that they belong to the Mallow family but none is fully hardy in Britain.

ACER is the Latin name for the 200-strong Maple family, mostly trees, and means 'sharp', but whether from the five-pointed leaves or because the wood was used for lances is uncertain. They come from many parts of the Northern Hemisphere but the Chinese and Japanese ones are the most decorative. The root meaning of 'Maple' in the Germanic languages seems to be 'food', perhaps because the young leaves were once eaten pickled.

AMELANCHIER is the Provençal name for the Snowy-Mespilus which in turn was Theophrastus's name for the Medlar. There are several species which make splendid hardy small trees for the garden.

AZALEAS are now included in the same genus as Rhododendrons which belong, like Heathers, to the great Erica family. They are wholly or partly deciduous. The name, from Greek, means 'dry' from the sort of situation they will tolerate and you may have noticed that they suffer less in a dry summer than the rest of the Rhododendrons.

BERBERIS as a name comes from Barbary, the old name for North Africa, where the Greeks and Romans thought bearded, uncivilised people or 'barbarians' lived and, indeed, pirates from the Barbary Coast were a problem for many centuries. The Barberries are certainly mostly prickly customers. *B. darwinii* with its orange flowers and holly-like leaves was discovered by Charles Darwin in Chile in 1835.

BOX. Buxus sempervirens (evergreen) is Virgil's name for the common Box, native on chalk downs in Southern England and much used since Tudor times to edge formal beds. Boxes were originally made from Box-wood and used for medicaments.

CALLUNA is the name of the Common Heather or Ling. There is only one species, but hundreds of named varieties for gardeners to choose from. The name comes from Latin *kalluno*, to cleanse, because heather, like Birch twigs, can be made into an excellent besom for sweeping clean.

CARYOPTERIS with its aromatic grey foliage and blue flowers comes from Asia. The name is from Greek *karyon*, ' nut' and *pteron*, 'winged'.

CEANOTHUS comes from the Greek for a spiny bush, but the blue-flowered ones we grow are not conspicuously armed with thorns. They are North American shrubs.

CHAENOMELES. The name of the ornamental Japanese Quince (often just called 'Japonica') is from Greek meaning 'gaping apple'. The true Quince, *Cydonia* (from Cydon in Crete) often does split and gape as it ripens in my garden, but perhaps the deep-set blossom end of the ornamental Quince gives rise to this curious name.

CISTUS with their single rose-like flowers are Mediterranean shrubs and not really hardy in my sodden and unkind corner of the Midlands. The name is the Greek one for the plant and means 'capsule' implying that it has conspicuous seed pods.

CORNUS, Cornel, Dogwood. What a big and useful family with flower, fruit, leaf and stem colour to recommend its various members. The Latin name is that for *C. mas*, the Cornelian Cherry.

Parkinson says our native Dogwood (*C. sanguinea*) has berries "not fit to be eaten, or to be given to a dogge" but before that it was Dagwood because it was used for making dags or skewers (c.f. dagger).

CORYLUS, Hazel. There is a purple-leaved form of *C. maxima*, the Kentish Cob, but the corkscrew form of our native *C. avellana* must be one of the most eye-catching shrubs in winter. *Corylus* is Greek and meant 'helmet' from the shape of the nut, while Hazel is an ancient name in the Germanic languages.

COTINUS was the Greek name for the Wild Olive. *C. coggygria* was formerly included in the Rhus genus. Its English nicknames, Wig Tree and Smoke Bush need no explanation to anyone who has seen its curious flowers. Coggygria is an ancient Greek name for Cotinus.

COTONEASTER means 'Quince-like' in Latin, though the resemblance is not obvious with the evergreen ones which are most useful to gardeners. *C. horizontalis*, which is good for hiding drain-covers, has been well nicknamed the Herringbone Bush.

CYTISUS is the Greek name for a clover-like plant and these Brooms and those in the closely related Genista genus all belong, like Clover, to the great Pea family (*Leguminosae*).

ELAEAGNUS is a total misnomer for the familiar variegated forms of *E. pungens* or, indeed, for any of the other 45 species you are likely to meet in a British garden. The name comes from two Greek words, *helodes* meaning 'growing in marshes' and *hagnos* meaning 'pure' or 'chaste' and evidently applied originally to a willow and its mass of white feathery seeds.

EMBOTHRIUM, Chilean Fire Bush. They come from W. Australia as well as S. America and are not fully hardy here. The Greek name refers to the way the anthers grow 'in small pits'.

ERICA is the name Pliny used for the Tree Heathers of the Mediterranean. We have half a dozen native Heathers in Britain and the Bilberry and Strawberry Tree are members of the same family but there are over 500 species world-wide.

EUCRYPHIA with their exquisite golden-stamened white flowers are named from the Greek for 'well-covered' because of the brownish sepals which protect the flower bud.

EUONYMUS is Theophrastus's name and means 'famed'. The deciduous ones like our native Spindle Tree (*E. europaeus*) are grown for their autumn colour or curious fruits while the evergreen ones are mainly valuable for their variegated foliage. 'Spindle' does not need explanation: it was adapted by Turner from German *spinnelbaum* in the sixteenth century.

FATSIA. There is only one Fatsia, in other words it is a 'monotypic' genus: it and its name come from Japan, so it is not surprising that its specific name is *japonica*.

GENISTA is Latin, the name found in Virgil. His *planta genista* gives rise to the name Plantagenet for our kings from Henry II to Richard III, since a sprig of golden Broom was the crest of the Angevins.

HAMAMELIS, Witch Hazel. In the old herbals you will find that Hornbeam was called Wych Hazel and, as with Wych Elm, it has nothing to do with witchcraft, but comes from an old word meaning 'pliable'. That was before the winter-flowering Hazels were introduced from the Far East last century. 'Hamamelis' is a misnomer too, since the Greek name applied to a tree with pear-shaped fruits.

HEBE, in Greek mythology, was a goddess of youth, the daughter of Jupiter and Hera and wife of Hercules. The Romans called her Juvenatus. Until recently these New Zealand shrubs were lumped together with the herbaceous species as Veronicas, and may still be found under that name in some catalogues and most of the older gardening books.

HELIANTHEMUM is Greek for 'Sun-Flower'. It is the name of the Rock-Roses which belong to the Cistus family and are no relation of either Sunflowers or Roses.

HIBISCUS is the old Greek name for Mallow to which family they belong. They mostly need a warmer climate than ours, though you can get away with the so-called Hollyhock Bush (*H. syriacus*) in most places.

HYDRANGEAS come from the Himalayas, China, Japan and North America. The name means 'water-jar' (Greek *hydor* and *aggos*) from the cup-like shape of the seed capsule. Pity it does not sound better in English. We need a good vernacular name for a splendid genus.

HYPERICUM, St John's Wort. The cheerful gold flowers of these unfussy shrubs are said to have been hung over pictures to ward off evil spirits, thus *hyper* (above), *eikon* (picture). Some scholars think the second element is *ereike* (heath) however.

JUNIPERUS is the Romans' name for our native *J. communis*, which will occasionally reach ten feet and is one of only three British conifers. There are many splendid small and dwarf varieties for the garden, including columnar and prostrate types.

LAVANDULA, Lavender, may have been introduced from the Mediterranean region as long ago as Roman times. It was originally grown for medicinal rather than decorative reasons and used for headaches, what Gerard calls "panting and passion of the hart" and less elevated ailments. The name comes ultimately from Latin *lavare*, to wash, because it was also used to scent soap and bathwater. There are several species, hybrids and varieties named after the places where they were raised like 'Hidcote' and 'Munstead', Gertrude Jekyll's garden. *L. spica* (with a long flower spike), also known as *officinalis* and *angustifolia* (narrow leaved), is the Old English Lavender and makes the largest bush. The 'Dutch' Lavender, *L. vera* (true) may only be a more compact form of *spicata*. The 'French' Lavender, *L. stoechas* (from the Iles d'Hyéres near Toulon, formerly called the Stoechades), was once the most highly prized by apothecaries, but is not so often seen today.

LEPTOSPERMUM means 'narrow-seed'. These evergreen shrubs have Heather-like leaves but are related to Myrtle. *L. scoparium* (broom-like) from New Zealand is known as the Manuka or Tea Tree and will grow ten feet tall in a sheltered place. *L. humifusum* (sprawling) is prostrate, nearly hardy and comes from Tasmania.

MYRTUS, Myrtle, is a Mediterranean shrub not unlike Bay, though the flowers are showier and fragrant, as are the leaves when bruised. It will only survive outdoors in mild districts and with shelter. The name comes via Latin from the Greeks who, in turn, borrowed it from Semitic, so it is related to the Myrrh of the Bible and meant 'bitter'.

OSMANTHUS. -anthus you will recognise as from the Greek for flower and *osme* is 'fragrance'. I remember searching for the source of a wonderful scent at Chirk one Spring and tracing it to the inconspicuous white flowers of *O. delavayi* on the castle wall. Abbé Delavay found it in China about 1890.

PHILADELPHUS, Mock Orange. Introduced late in the seventeenth century to England, the flowers were thought to resemble orange-blossom. See the introduction to this chapter for the scientific name-change.

PIERIS is the 'surname' of the Muses who were the Goddesses of the Arts: they were the nine daughters of King Pierus of Emathia in Macedonia and so they were known as the Pierides.

PITTOSPORUM means 'pitch-seed' from their resinous covering. You may never have seen them growing since they are only happy in the mildest parts of the country but you will know the bright green leaves on black twigs of *P. tenuifolium* which florists frequently add to bouquets of cut flowers.

POTENTILLA means 'a bit powerful' from Latin *potens*. I do not know whether the shrubby ones were used medicinally but Culpeper is full of the virtues of our native Cinquefoil for agues, infections and inflamations.

PRUNUS is the Latin name for the Plum Tree. Besides the evergreen Portugal and Cherry Laurels mentioned at the beginning of the chapter, the genus of 430 species includes all the edible cherries, almonds, gages, sloes, etc. as well as many ornamental trees and shrubs.

PYRACANTHA. The English name 'Firethorn' is an exact translation of the Greek. Some authors suggest that the 'fire' refers to the smarting caused by the thorns rather than to the colour of the berries.

RHODODENDRON is the Greek for 'red' or 'rosy tree' and was originally the name of an Oleander. There are 500 of them (including the Azaleas): Few will tolerate lime and all flower in spring or early summer, the only faults of a genus full of marvels.

RHUS is the old Greek name for the Sumach which in turn comes via French from a Syriac word for 'red' since *R. coriaria* produces a red dye. The familiar *R. typhina* is happily nicknamed the Stag's Horn Sumach; it ought to produce a yellow dye to judge from the colour of the under-bark.

ROSA. Old English borrowed Latin *rosa* which is related to Greek *rhodon* which seems to have meant what we should call rose-red, a colour. It therefore seems a pity that yellow Rockrose and white Christmas Rose should be so called. Today, the five-petalled dish shape of the single Rose is not most people's first association with the name.

SAMBUCUS, Elder. Our native Elder is almost a weed, it springs up so readily wherever birds drop its seeds but there are more decorative species, especially the golden *S. racemosa Plumosa Aurea* and *S. racemosa* with scarlet berries, which you can find naturalised in the vicinity of Victorian shrubberies. The Old English name was Ellern and Sambucus was the Latin and Greek name.

SANTOLINA, Cotton Lavender. The name of these aromatic natives of the Mediterranean region comes from Latin *sanctum linum* meaning 'holy flax'.

SKIMMIAS come from Japan and elsewhere in the Far East. The name is an attempt at the Japanese *Miyami-Shikimi*.

SORBUS is the Latin name for *S. domestica*, the Service Tree whose fruits were known as sorb-apples. It is probably a native, as are Rowan and White Beam, and was known to the Anglo-Saxons as *syrftreow* which also comes from Latin *sorbus*.

SPIRAEA. There are a lot of these hardy, deciduous shrubs whose name comes from the Greek of Theophrastus for a plant used in making garlands. Happily, therefore, one of the best white-flowered hybrids is called 'Bridal Wreath'.

TAMARIX, Tamarisk. They do not like lime but will put up with sand and salt spray. The name is the Latin one which may come from the area of the Spanish River Tambo which the Romans called Tamaris or perhaps it is related to Hebrew *tamar*, 'palm'.

THUJA (or Thuya). These useful, slow-growing small conifers have the name which Theophrastus gave to a resinous tree with aromatic wood which is sometimes assumed to be the Juniper.

VIBURNUM. This big genus which contains many splendid

shrubs takes its name from the Latin for our native *V. lantana*. The English name, Wayfaring Tree, seems to have been coined by Gerard who thought the Old French *viorne* came from *via* and *ornata*, a tree that ornaments the road or way. Was it, however, a tree for wayfarers as Geoffrey Grigson implies, or was Gerard thinking of 'fairings', the ornaments or trinkets one bought as presents from pedlars at fairs? (See Autolycus in Shakespeare's *The Winter's Tale* IV, iii.)

ZENOBIA. There is only one Zenobia and it is called *pulverulenta* because of its powdery-looking leaves. It is one of the Ericas, comes from the S.E. United States and was named after the ambitious Queen of Palmyra who aspired to be known as 'Queen of the East' until she was defeated by the Romans in AD 273.

Chapter 9
A Clutch Of Climbers

Why do we talk of a *clutch* of eggs? It seems that it has nothing to do with the number we could grasp in a handful nor yet with the clucking of a broody hen: it comes from an obsolete verb meaning to hatch. The verb to clutch is quite unrelated but suggests to me a useful collective noun for climbing plants. They all clutch hold of their neighbours or some support in the upward struggle towards the light. We may think of plants as competing mainly for moisture and nutrients with their roots but, in a climate like ours, the need to get the maximum amount of light for their leaves is just as powerful an imperative. Indeed, a speeded-up film of plants growing in competition with each other shows a blind struggle every bit as fierce and ruthless as anything the zoologist can show us of nature red in tooth and claw.

A quick tour round the garden will reveal something of the variety of ingenious methods plants have developed for getting one up on the neighbours. You may have disciplined them a bit, but their jungle instincts are easily seen. There are stickers and clingers, smotherers and scramblers, twirlers and stranglers.

Least sophisticated in their adaptation are the scramblers like Blackberries and Rambler Roses which shove and grapple their way up through shrubs and prevent themselves from being dislodged with thorns. If we deprive them of natural support, we have to tie them up. Cleavers or Goose Grass uses the same method of clinging, though its downward-pointing hooks are microscopic.

Then there are the Ivies with aerial roots which enable them to grip rock-face, brickwork or tree-bark. They do not feed off trees parasitically, as is sometimes supposed, but merely use them as a climbing-frame. Even more remarkably equipped is the Virginia Creeper whose short tendrils are tipped with sticky pads which enable it to climb a completely smooth surface such as a window pane. Their grip is powerful enough for chunks of plaster or mortar and not merely flakes of paint to be dislodged when the creeper is pulled from the wall on which it has fastened itself.

Next come the twiners which encircle their victims and climb up them spirally. See the young shoot of Bindweed or Runner Bean sweeping round in a circle until it touches something it can lasso. Everyone who remembers the song by Michael Flanders and Donald Swann will know that the Honeysuckle climbs clockwise and the Bindweed anti-clockwise. Hop and Runner Bean would make a similar left- and right-handed misalliance: the Hop has hair-like hooks on its stem which help it to cling but the Runner Bean does not, and will need some encouragement to get started on a smooth surface like a garden cane. Once it gets going, however, it can complete a revolution in three to four hours. There are other twisters which will climb either way depending on which side of the growing tip first makes contact with support.

Try growing one of these climbers in a pot and then play a dirty trick on it: turn the pot upside down and see what happens. In fact it will climb back up itself in the opposite direction to the force of gravity. This is called 'negative geotropism'. Some of them can also perform a stunt or two of their own: you will often see shoots of Black Bryony or Hop twisting around each other and managing to rise a yard or more above the top of a hedge, like the Indian rope-trick, in their search for higher things. Ivy, when it over-tops a tree or wall, will become self-supporting. The leaves then lose their typical lobed shape and flowers are produced. If you take a cutting from this part of the plant and root it, you can grow it on as a 'tree-Ivy' without support, a trick that Victorian gardeners were fond of.

Finally, and most remarkably, there are plants which have developed arms, that is to say specially adapted leaves, to help them to hold on while climbing. With Clematis and Canary Creeper it is the leaf-stalk or petiole which twists around a supporting twig. With Sweat Pea, as with wild Vetches, some leaves have developed into thread-like tendrils which very effectively lash it onto anything they touch. Some tendril-using climbers, like the Cobaea or Chilean Bell-Flower show an additional sophistication: once the tendril has encircled something solid, it contracts like a spring giving support which is firm but elastic.

Some genera contain one or two climbing species though they

are better known for their non-climbing members - Gazania, Pelargonium and Hydrangea would be examples - and they are not generally included in this chapter. Neither are those shrubs like Chaenomeles which are usually grown against a wall but are not really climbers. Here now is a clutch of others which seem, like squirrels, cats and small boys, to have climbing in their blood.

ARISTOLOCHIA, Dutchman's Pipe, Birthwort. There is a Mediterranean species, *A. altissima*, but the hardy one, *A. durior or macrophylla* is N. American and makes an interesting wall covering with its broad heart-shaped leaves and extraordinary green-yellow flowers striped with chocolate. The greenhouse species, *A. elegans*, makes me think of a saxophone or tuba rather than a tobacco pipe. The generic name from Greek translates as 'best childbirth' but the authorities I have consulted are divided on whether the plants were supposed to prevent or procure abortion.

BIGNONIA is a family named after the Abbé Jean Paul Bignon (1662-1743) who was librarian to Louis XIV of France. It has been much rearranged recently. The climber in the family most commonly grown for its vigour and hardiness is the Trumpet Creeper or Cross Vine, *Campsis radicans*, with orange-scarlet flowers. *Campsis* comes from Greek for 'bent' from the shape of the stamens.

BOUGAINVILLEA. It is the papery bracts of brilliant rose, magenta or orange-scarlet rather than the insignificant flowers which make these tropical vines so showy. They are named after LouistAntoine de Bougainville (1729-1811), explorer, mathematician, scientist, lawyer, who sailed around the world, 1767-9, and was made a Fellow of the Royal Society in London.

CAMPSIS, Trumpet Creeper, see Bignonia.

CLEMATIS. There are 250 species and plenty of hardy ones to choose from for the garden. They are all climbers though I have seen the lovely *C. alpina* making a clump like a rock-plant 4000 feet up in the Alps. The name is Greek and comes from *klema*, a vine-shoot. It had already been adopted for our native Old Man's Beard in the seventeenth century.

CLIANTHUS, Parrot's Bill. The name means 'glory-flower'.

C. puniceus (crimson) from New Zealand can be grown out of doors in mild districts. The only other species, from Australia, is not hardy and is prostrate rather than a climber.

COBAEA, Chilean Bell Flower, Cup-and-Saucer Plant. *C. scandens* (climbing) is the most commonly grown as a vigorous annual climber. Its rather sombre purple flower has suggested another name, Cathedral Bell. The genus was named after a Spanish missionary to S. America, Bernado Cobo (1572-1659).

ECCREMOCARPUS climbs like Cobaea by using leaf-tendrils but it is evergreen, has tubular orange-scarlet flowers and is not hardy in most districts. It is sometimes called Chilean Glory Flower (see Clianthus) but *Eccremocarpus* means 'hanging fruit'.

GARRYA, Silk Tassel Bush. *G. elliptica* from California and Oregon is grown for its decorative long grey-green male catkins. It is one of the plants brought back from the Columbia River by David Douglas (see Ch. 4) and named after Nicholas Garry, Deputy Governor of the Hudson's Bay Company between 1822 and 1835, who helped to make Douglas's expeditions possible.

HEDERA is the Latin name for Ivy which, in turn, comes from Old English. I like the bluntness of the Dutch name *klimop*, 'climb up'. There are fifteen species, some excellent as pot-plants and others, like the large-leaved Canary and Persian Ivies for smothering architectural eye-sores. There are also cheerful gold and silver variegated forms of our common *H. helix* which has the great virtue of producing flowers in October and berries in March, thus keeping bees happy at the end of one season and thrushes at the beginning of the next. There is nothing that provides better cover for roosting and nesting birds and hibernating insects. Conservationists should encourage it whatever arboriculturalists say.

IPOMOEA, Morning Glory, was introduced as long ago as 1629 and once known as Indian Bindweed. *Ipomoea* is from Greek and means 'worm-like'. There are 500 mostly tender species including yellow, white and scarlet flowering ones as well as the most familiar *violacea* from Mexico.

JASMINUM. There are 300 species, several worth considering as well as the indispensable (but straggly) Winter Jasmine, *J.*

nudiflorum and the vigorous, fragrant *J. officinalis* of summer. The name is neither Latin nor Greek for once but comes from Persian *Yasmin*.

LATHYRUS, Sweet Pea. *Lathyrus* is simply Greek for pea. The annual species which came from Italy in the eighteenth century is rightly called *odoratus*. Perennial ones like *latifolius*, the familiar Everlasting Pea of cottage gardens, are not fragrant and can be difficult to eradicate if you get them in the wrong place.

LONICERA, Honeysuckle. There are 200 species ranging from the familiar scented hedgerow climber *L. periclymenum* (twining around) to the useful hedging shrub, *L. nitida* (glossy) which has only once produced unmistakeable tiny Honeysuckle flowers and blue berries in my garden, and that was after the hot summer of 1976. Beware, 'Honeysuckle' is often the name given to Red Clover in old books. The genus is named after Adam Lonitzer (1528-1586), a German botanist whose *Kreuterbuch* was reprinted many times.

PARTHENOCISSUS, Virginia Creeper. *Kissos* is Greek for Ivy and *partheno* is virgin so the scientific name has been made up from the common one. It used to be considered a vine (Vitis) rather than an Ivy, but now has its own genus containing ten species. Some are native to China and the Himalayas while the familiar *quinquefolia*, of course, is N. American. They make up for insignificant flowers with fiery autumn foliage.

PASSIFLORA. Passion Flower, is so named because, when it was first found in tropical America, the flowers were thought to contain the symbols of Christ's passion, including the Apostles and the crown of thorns. There are in fact 500 species of vigorous but tender climbers. *P. edulis* produces the edible egg-shaped purple or yellow fruits.

PLUMBAGO, Leadwort. Pliny thought the flowers looked leaden, hence the Latin name from *plumbum*. He couldn't have known the two species in cultivation, *P. capensis* (from the Cape, S. Africa) with light blue flowers or *P. Rosea* from the E. Indies with scarlet ones.

POLYGONUM is a large genus featuring elsewhere in this book, the name meaning 'many joints'. *P. baldschuanicum* has to be included here because it is such a well known smotherer. It came from Baldschuan, near Bokhara in the Soviet Union, hence its common name of Russian Vine. It can grow up to 15 feet in a season and 40 feet overall and looks splendid where there is room for it to froth happily.

SOLANUM is another huge family which bears Pliny's name meaning 'comforter'. *S. crispum* (with curled edges) is a 15-20 foot scrambler known as the Chilean Potato-Tree. It has star-shaped mauve flowers with prominent golden anthers in the centre which do look very like those of some potatoes (which are also *Solanaceae* of course). There is another climber, *S. jasminoides*, known as Jasmine Nightshade, which is more tender.

STEPHANOTIS, Wax Flower, Madagascar Jasmine. *S. floribunda* will grow to ten feet in a greenhouse but is difficult to keep as a houseplant because it needs a humid atmosphere. The scent of the white flowers is irresistible. The name is the Greek one for Myrtle which was used to make a *stephanos*, or coronet, plus *otos* which means 'ear': you would have to pull the flower apart to see the ear-like parts of the crown of stamens.

THUNBERGIA. There are 200 species but the ones in cultivation are mostly tender climbers, *T. alata* (winged), Black-Eyed Susan, being the most familiar. Carl Peter Thunberg (1743-1828) was a pupil of Linnaeus who joined the Dutch East India Company as a doctor and sent back plants to Sweden from Japan and Africa.

TRACHELOSPERMUM. *T. jasminoides* is the most often grown, usually as a wall climber. It is an evergreen but the flowers do resemble those of White Jasmine. The name means 'neck-seed' apparently from their narrowness.

TROPAEOLUM, Nasturtium, Canary Creeper. There are 90 species, not all of them climbers like *T. peregrinum*, the Canary Creeper, which can grow twelve feet in a season. There are also perennials like *T. speciosum*, the Flame Creeper. The name *tropaeolum* means 'trophy': Linnaeus chose it when he saw Nasturtiums climbing up a post because it reminded him of the way the Greeks hung the trophies of battle, the shields and helmets of their fallen enemies, on a pole or tree-trunk.

VITIS, Grape Vine. *Vitis* is simply the Latin name for the Grape Vine and our words 'vine' and 'wine' ultimately come from the same root. There are about 70 species and they are true climbers, clinging by tendrils. There are several species, especially *V. coignetiae* (named after a Mme Coignet who collected the seeds in Japan), which is known as Crimson Glory Vine, that are grown for the splendid colours their heart-shaped leaves turn in autumn.

WISTERIA is a misspelling since the genus was named after Caspar Wistar (1761-1818) who was a professor of anatomy at the University of Pennsylvania and one of the owners of what is now Vernon Park. These climbers, however, are not American but come from China and Japan where, at Ushijima, there is a famous specimen with a trunk which is 32 feet round and bearing an estimated 80,000 trusses of flowers.

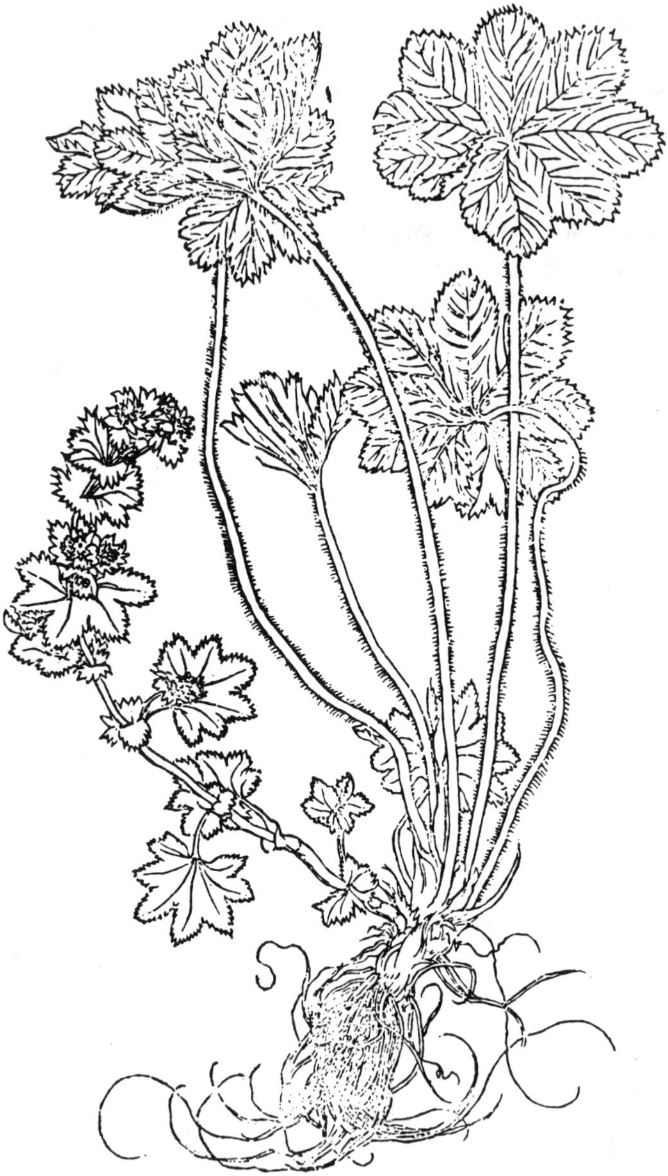

Lady's Mantle

Alchemilla vulgaris from Otto Brunfel's *Herbarum*, 1530.

CHAPTER 10

A GRABBLE OF GROUND COVER

To cultivate soil and leave any part of it unoccupied is to invite in colonising weeds and the problem of controlling them, a chore most of us can do without. In a sense the ultimate ground cover plant is grass but it must be mown throughout the growing season and, even with a strimmer, is time-consuming to keep tidy around trees and shrubs. What we need, ideally, is an evergreen, ground-covering, weed-smothering plant which is attractive in its own right, unfussy about soil or situation and fast-growing but not too aggressive. In the real world, no such paragon of all the virtues actually exists, but there are quite a lot of plants with many of them, and they are often not as well known as they might be.

A selection will be found in this chapter, but there are very many solutions to ground cover problems. Climbers like the Ivies need little encouragement to crawl and annuals like Nasturtiums can be very effective for a season. There are excellent prostrate shrubs, Junipers, Cotoneasters, Willows and St John's Worts, a Hebe like 'Pagei' or a Viburnum such as *davidii*, and Heathers, of course. Very many rock-plants are spreaders which soon become too vigorous for the small rockery but make pretty weed-proof mats and many herbaceous species have colonising vices in the border which become virtues when they are translated to a different situation. It is all a question of the size and site of the ground to be covered.

To talk of a 'grovel' of ground cover plants would be insulting, implying as it does crawling face-downwards in fear or humility. However, we borrowed a verb 'to grabble' from the Dutch in the sixteenth century which can mean 'to sprawl and clutch the ground' with no pejorative undertones. A 'grabble' let it be then, and when they have sprawled and spread happily as nature intended, they may be a 'grandeur' and a 'gratification' as well.

ACAENA, New Zealand Burr, Creeping Burnet, Biddy Bid. There are 100 species, one of which, *A. anserinifolia* (goose-like leaves), has become naturalised in some places in this country, the seeds having arrived with Australian wool. Three useful ones for the garden are *A. microphylla* (small-leaved) with bronze-green foliage, *A. caesiglauca* (lavender-grey) and *A. buchananii* which is grey-green and was named after John Buchanan (1819-1898), who was botanist with the Geological Survey in New Zealand. They all make a dense mat of stems and leaves and have inconspicuous flowers followed by decorative burrs. Watch out for your socks, though, and I once had to comb a mass of them out of the fur of an incautious cat. Greek *akaina* which gives the generic name is a thorn or prickle.

AJUGA, Bugle. There are good bronze, purple and variegated varieties of *A. reptans* (creeping) and *A. pyramidalis*, both native species which like moist soil but the names are a puzzle. 'Bugle' comes from the late Latin name *bugula* by which it was known to the physicians and apothecaries. I suppose that, like Bugloss and the musical instrument, it is related to Latin *bos*, an ox. As to *Ajuga*, the authorities say variously that it is obscure, that it comes from Greek *a*, 'not', and *zugon*, 'a yoke' (but why?) and that it is a corruption of Latin for an abortifacient. Culpeper mentions many ailments it will help or cure, including nightmares, but says nothing about abortions.

ALCHEMILLA, Lady's Mantle. The name may be latinized from the Arabic *Alkemelych*, 'the alchemist's plant' and one story is that they collected the pearls of dew which sit on the backs of the leaves for use in their experiments to find the Philosopher's Stone. They were also reputed to restore faded female beauty. Both *A. mollis* (softly-hairy) with its foamy lime-green flowers and the smaller *A. conjuncta* (the beautiful, silky leaflets are joined together at the base) are closely related to native species.

ARISARUM, Mouse-tail Plant. Dioscorides called a tuberous herb *arisaron* which is thought to have been *A. vulgare*. However, *A. proboscidium* (like an elephant's trunk) will make good ground cover under trees. Beneath the shiny green spear-shaped leaves, the Arum-like flowers have spadices which suggest mouse-tails rather than the proboscises of pachyderms.

ASARUM europaeum, Wild Ginger, also known as Asarabacca, is a rare and curious native woodland plant with glossy kidney-shaped leaves and small brownish-purple flowers which open at ground level and are usually hidden in the leaf-litter. A similar American species, *A. caudatum*, will be quicker to make a carpet under trees. 'Asarum' is Dioscorides' name and it is also called 'Bacc(h)ar' by the herbalists but whether from *bacca*, a berry, or Bacchus seems uncertain: they used it as a powerful purgative.

ASPERULA, Woodruff. Our native Sweet Woodruff which makes attractive ground cover is now classed as one of the Bedstraws and called *Galium odoratum* while our only *Asperula* is Squinancy-wort, *A. cynanchica* (of quinsy, literally 'dog-throttling'). Other Asperulas are best thought of as rock-plants. 'Asperula' means 'rough' and Woodruff, like several of its relatives, has hooked hairs on the edges of the leaves which are arranged in a whorl suggesting also a 'ruff'.

BERGENIAS are named after a professor of Frankfurt an der Oder, Karl August von Bergen (1704-1760) but were formerly classed as Saxifrages or Megaseas. There are several species but most garden varieties are hybrids with *B. cordifolia* (with heart-shaped leaves) giving quite a range of flower colour. They are tough customers and I have known their leaves vanquish Squitch and Bindweed.

BRUNNERA macrophylla (large-leaved) used to be called *Anchusa myosotidiflora* and the blue flowers are certainly Forget-me-not-like. There is an attractive variegated form. The genus is named after the Swiss botanist, Samuel Brunner (1790-1844), though actually discovered growing in woods near Tiflis by an Englishman, Dr J. Adams.

CERASTIUM, Snow in Summer, is related to Chickweed with a similar talent for rapid spreading. *C. tomentosum* (thickly matted with hairs) and *C. biebersteinii* (after a German botanist of that name) are very similar though the latter is the more woolly and silver, and forms the denser ever-grey mat. The name comes from Greek *keras*, a horn, from the shape of the seed capsule.

CHRYSOGONUM virginianum from N. America does well in damp shade. The name means 'golden knee' because the star-like flowers spring from the joints of the stem.

CONVALLARIA, Lily of the Valley. There is only one species,

C. majalis (of May) and *convallium* is Latin for a deep valley. I find them temperamental, sometimes thriving in the most unpromising places and their rhizomes capable of pushing up paving stones. In the wild they seem to like shade and leaf-mould over alkaline rocks.

EPIMEDIUM, Barrenwort, Bishop's Hat. The name was used by Dioscorides and Pliny but for quite different species and it appears to mean something like 'on or above average (height)'. The best species are Japanese, the evergreen ones making useful ground cover and the deciduous ones have such colourful foliage that they deserve a place at the front of the border.

GALIUM, see *ASPERULA*.

GAULTHERIA procumbens, Partridge Berry, Checkerberry, Creeping Wintergreen from N.E. America is excellent on lime-free soil with a carpet of shiny, dark green leaves, white flowers and bright red berries. It is a big genus named after Jean Francois Gaulthier, a Canadian physician-botanist who died in 1858. There are bushier species like *G. shallon* (the American Indian name) with purple berries much loved by pheasants, and neat rockery subjects like *G. sinensis* from Burma.

HEUCHERELLA is the name coined for intergeneric hybrids between *Heuchera* (see p.40,) and the Foam Flower, *Tiarella* (see p.115). They make better spreaders than either parent, *H. tiarelloides* having salmon flowers and *H. alba* usually tinged with pink.

HOSTA, Plantain Lily. Once known as Funkias, they come from China and Japan and were named after Nicholaus Tomas Host (1761-1834), an Austrian botanist who was Physician to the Austrian Emperor.

HOUTTUYNIA is a genus with a single species, *cordata*, from the Far East and named after a Dutch naturalist, Martin Houttuyn (1720-1794). Its underground runners spread rapidly in damp soil, the heart-shaped (cordate) leaves have an orangey smell and may be maroon-tinged and the white 'flowers' are actually bracts.

HYLOMECON from Japan has a creeping rootstock, ferny leaves and yellow, four-petalled flowers like a Greater Celandine. The name means 'wood-poppy' and it likes moist wood-soil.

IBERIS, Candytuft, comes from Iberia, Spain, though we have a native, pink-flowered annual *I. amara* (bitter) in chalky fields in Southern England. It is the perennial sub-shrubs, however, like *I. sempervirens* (evergreen) with white flowers which are invaluable for smothering a dry bank where they always manage to look bright even in dirty situations like roadsides. 'Candytuft' is from 'Candy', an old name for Crete, another home of the genus.

LAMIUM comes from Greek *lamios*, 'throat' from the shape of the corolla. The Dead nettle to go for is *L. maculatum* (spotted) with silver-striped leaves and pink-purple flowers. The variety 'Beacon silver' is particularly attractive. A more rampant spreader, even in heavy shade, is Yellow Archangel *L. galeobdolon* (a name used by Pliny meaning 'weasel-smell'). The variety 'Variegatum' has handsome black and silver markings on the leaves.

LITHOSPERMUM. Best known is the 'Heavenly blue' variety of *L. diffusum* (spreading), and there are few truer blues but it does need well-drained, lime-free soil and will not stand much shade. The name means 'stone-seed' from their hardness.

MAIANTHEMUM bifolium easily translates as 'two-leaved Mayflower' and, though deciduous, it will make a good carpet given shade and moisture. It is known as the May Lily in this country where it is a rare native in Yorkshire and Lincolnshire, and as the False Lily-of-the-Valley in the U.S.A.

NEPETA, Catmint. Our native *N. cataria* (of cats) was grown in herb gardens before the introduction of the more decorative *N. mussinii* from the Caucasus which was named after Count Mussin-Puschkin. This is now properly called *N. faassenii* after the Dutch nurseryman who produced the hybrid we actually grow today. 'Nepeta' is an old name said to be from a place called Nepet in Tuscany which I cannot find in my atlas. It is the visiting toms, not our resident moggy, who show any interest in it.

OMPHALODES comes from Greek *omphalos*, a navel, because of a suggestive depression on the seed, hence the English name Navelwort. The other names, Blue-Eyed Mary and Rock Forget-me-not give a good idea of the flowers. Those of *O. verna* appear in March and have a white eye; *O. cappadocia*'s are clear blue and a bit later. Both combine particularly happily with Narcissi in a woodland setting.

PHLOMIS, Jerusalem Sage. There are many species and their woolly foliage suggested the name from Greek *phlogmos*, a wick. *P. fruticosa* (shrubby), from the Mediterranean, is evergreen with whorls of yellow flowers and needs sun. It will sprawl vigorously among shrubs or the taller herbaceous plants.

PRUNELLA. Our native Self-heal is *P. vulgaris*. *P. grandiflora* is showier with the same purple colour of flower. Best of all are the 'Loveliness' varieties of the hybrid *P.x. webbiana* in shades of pink, lilac and white. As the name implies, the plant was used to cleanse and heal wounds and sores and was known to the herbalists as Brunella or Prunella from Latin *prunum*, plum-purple.

PULMONARIA, Lungwort, Jerusalem Cowslip, Spotted Dog. *P. longifolia* is a rare native while the much commoner *P. officinalis*, sometimes also called Soldiers and Sailors, is an escape from herb gardens. It was used to make linctuses for bronchial ailments, a classic case of the Doctrine of Signatures in action since the blotched leaves were thought to look like lungs, *pulmones* in Latin. *P. saccharata* (sugary) from Italy has the most striking silver-spotted leaves while those of *P. angustifolia* (narrow-leaved) are plain but the flowers are sky-blue.

STACHYS comes from Greek for a spike because of the shape of the inflorescence. We have several native Woundworts and Betony, *Betonica officinalis*, which deserves a place in the garden, used to be *Stachys betonica*. The best known member of the genus is *S. lanata* (woolly), now properly called *olympica* after Mount Olympus, and variously known in English as Lamb's-Tongues, -Tails or -Ears. *S. macrantha* (large-flowered) is also mat-forming and has an abundance of spikes of rose-purple flowers but the leaves are green and deciduous.

SYMPHYTUM comes from Greek *symphysis*, 'growing together', and the English name, 'Comfrey', comes via French from Latin *conferre* with the same meaning. *C. officinalis* had many medicinal virtues and because its parsnip-like root yields a starchy substance which was used like Plaster of Paris it was also known as Knit-Bone or Bone-Set. The species used under trees and shrubs is the ivory-flowered *S. grandiflorum* from the Caucasus which spreads rapidly and is only about nine inches high.

TIARELLA, Foam Flower, False Mitrewort. *T. cordifolia* (heart-shaped leaves) from N. America is the one known as Foam Flower. Actually the leaves are more Maple-like and yellow-green and there are other species with Ivy-shaped leaves like *T. trifoliata* but they all have spikes of foamy white or pink-tinged flowers. *Tiarella* is a little tiara or diadem from the form of the seed-pod.

TRACHYSTEMON orientalis has spikes of blue flowers in Spring followed by large weed-smothering leaves and is also useful because it tolerates dry shade. Tiresomely it has several aliases such as *Normannia cordifolia* and *Borago* or *Psilostemon orientalis*. The name comes from the rough (*trachys*) stamen of one species.

TRILLIUMS used to be classed as Lilies but now have a family of their own. There are 30 species from the U.S.A. and Canada, about half a dozen of which are cultivated as attractive ground cover for a moist shady situation. The leaves, sepals and petals are all conspicuously arranged in threes which accounts for the name.

VINCA is from the Latin verb *Vincio*, to bind, and 'Periwinkle' comes from the same source, *pervincio*, to bind about, by way of Norman French *pervenke*. They are stragglers really and perhaps best allowed to trail through other ground cover plants. *V. major* regularly climbs six feet up my *Clematis montana* to good effect. There are forms of *V. minor* with variegated leaves and purple, Burgundy and white flowers which can be double.

WALDSTEINIAS were named after an Austrian, the part-author of a beautiful work on the plants of Hungary, Count Franz Adam Waldstein-Wartenburg (1759-1823). There are half a dozen species but the only one I have seen is *ternata* (with three parts) which is also known as *trifolia* and the glossy evergreen leaves are very like those of a Strawberry. It will also spread like an Alpine Strawberry and is indifferent to soil or situation though in shade the *Potentilla*-like flowers will be fewer.

Lords - and - Ladies
Arum maculatum from Leonhart Fuchs' *Herbal* of 1542.

Chapter 11

A Recumbence Of Rock Plants

It would be tempting to talk of a 'romp' or a 'rollick', a 'razzle-dazzle' or even a 'rampage' of rock plants. Indeed, one way of dealing with a large bank is to plant it up with vigorous spreaders and let them fight it out among themselves. I confess that this is what has happened to part of my garden more by accident than design. It looks good and requires no attention. Many of the ground cover plants in another chapter would make suitable subjects for this approach. However, if you have a small area, want to grow a variety of beautiful things and have the additional pleasure of designing a landscape in miniature, the rompers must be strictly excluded though they are often sold as rock plants and look demure enough sitting in pots at the garden centre.

Most writers say that rock gardening, like so much else, began long ago in the Far East where rocks, trees and water were arranged in cunning imitation of nature at its most fantastic. A Scot called William Forsythe (after whom Forsythia is named) is credited with having created the first rockery in England, in the 18th century, using stones from the Tower of London, but I suspect this was a rather formal and regimented affair. The informal rock garden, a pseudo alpine landscape, is little more than a century old. The earliest use of the word 'rockery' recorded by the Oxford Dictionary is in 1845 and that was in connection with the Victorians' passion for collecting and growing ferns. When William Robinson published his book *Alpine Flowers for English Gardens* in 1870, the rock garden had become a popular feature.

Genuine alpines are generally recumbent rather than rollicking, they sit tight and keep a low profile with good reason: where they come from, to stick your head up more than an inch or two would be asking to have it blown off. Many of them will have specific names like *alpina* or *montana*, *nana*, 'dwarf', or *repens*, 'creeping'. Some will come from arid, stony places and want sunshine and sharp drainage; others may spend much of the year blanketed by snow and resent a combination of wet and cold; others again are at home on misty moors and must never dry out: they may be easier to grow in Scotland than in the S. of England. It always helps to know where your plants come from.

A rock garden is not strictly necessary: many alpines will grow happily in chippings or gravel between paving stones or in troughs or on a dry stone wall. If you *are* going to build a rockery, the golden rules are to use local stones, the larger the better, and to lay them as they would be exposed in a natural cliff or a cutting. The 'plum-pudding' style of rockery, a mound of soil with rocks stuck into it, never looks right. Plum-pudding is too rich a diet too; poor soil and good drainage will suit the great majority of alpine plants which would fail if it was too fertile and heavy. Many genera contain some alpine species, so this chapter is a gathering up of ones which have not appeared elsewhere. Do not be surprised, therefore, to find no Aubretia or Campanula, Dianthus or Gentian: their names, attached to larger relatives, can all be located through the index.

ACANTHOLIMON, Prickly Heath, Prickly Thrift. There are many of them and they come from dry hillsides from Greece eastwards, so thrive on sunshine. The name is from Greek *akantha*, a thorn or spike, and *limonium*, Sea Lavender, to which the genus is related. Most commonly grown is *A. glaumaceum* (from Latin *glauma*, a husk, because of the dry bracts).

AETHIONEM, Stone Cress, Candy Mustard, Lebanon Candytuft. Several species are grown, mostly with rose-pink flowers and all like sun and lime. The name seems to be from Greek *aitho*, to scorch, and *nema*, a thread, from the singed appearance of the filaments or, perhaps from *aethes* (unusual).

ALYSSUM, Gold Dust, Madwort. For a small rockery, look out for *A. montanum* or *A. alpestre* rather than the usual forms of *A. saxitile* (of rocky places). Somehow Alyssum got the reputation of curing madness, the name coming from *a lyssa*, 'without rage'.

ANACYCLUS, Mount Atlas Daisy, a name that tells you they like sunshine and good drainage. The underside of the petals (and therefore the bud) is red, opening into a yellow-white daisy. The flower disc is surrounded by a ring (Greek *kuklos*) of ovaries.

ANAGALIS, Pimpernel. As natives we have the Scarlet and the pink Bog Pimpernels. There is a cultivated form of the latter but it is the intense gentian-blue of *A. linifolia* (flax-like leaves) from the Mediterranean which gardeners usually want. 'Pimpernel' is a misnomer; it means 'little pepper' and seems to have been transferred by mistake in Victorian times from Burnet Saxifrage, *Pimpinella saxifraga*, which does have peppery leaves. 'Anagalis' is from a Greek verb to laugh, and the Pimpernels were anciently supposed to relieve sadness.

ANTENNARIA, Everlasting. The species grown have silver-grey foliage and pink or white flowers, but the name refers to the feathery hairs (pappus) on the seeds which resemble a moth's antenna, a Latin word which means, literally, the yard of a sailing ship.

ARABIS, Rock Cress. There are weeds and choice rock plants as well as the common and rather invasive *A. caucasia*. There is some muddle about the name; it ought to mean 'from Arabia' but the genus is not found there so possibly it implies that it will withstand drought.

ARENARIA, Sandwort. Our native Sandworts, Pearlworts, Stitchworts and Spurries are confusing near relatives and you will find the confusion extends to the cultivated rock plants. The best Arenarias (from Greek *arena* meaning sand) like *A. montana* make trim cover for dwarf spring bulbs before producing their own unspectacular white flowers.

ARMERIA, Thrift, Sea Pink. At one time it was known as 'Threepenny Bit Wort' from its appearance on the obverse of the twelve-sided George VI coin. Our native *A. maritima* is surprisingly happy far removed from sea cliffs and was popular for edging beds from Tudor times onwards. Other species like *A. caespitosa* (forming tussocks) from Spain need good drainage. 'Armeria' is the Latinised form of Old French *armoires*, the Sweet William, a Dianthus, of course, and quite unrelated but also having a clustered head.

CALANDRINA. The magenta Rock Purslane comes from the mountains of Chile and Peru but is named after a professor of Geneva, Jean Louis Calandrini (1703-1758).

CALCEOLARIA, Slipper Flower. The pouched flowers

suggested Roman half-shoes, *calceolus* in Latin. As well as the showy pot-plants from Mexico, there are dwarf alpine species from Patagonia like *C. darwinii* which need a cool moist situation.

CALLIRHOE, Poppy Mallow. In Greek mythology she was daughter of the river god Achelos. *C. involucrata* (surrounded by bracts), from California, is an excellent little rock plant for a dry spot.

CASSIOPE was the mother of the beautiful Andromeda who was chained to a rock and rescued by Perseus from the attentions of a hungry monster - a subject popular with artists. It is the name of a genus of the Heather family at home in the Arctic and on northern mountains. The only species I know is *C. lycopodioides*, (like Wolf's-foot, a Club Moss), which is mat-forming with single Heath-like white bells.

CODONOPSIS means 'like a bell' (Greek *kodon*) and they belong to the Campanula family. *C. ovata* (egg-shaped - the leaves) from the Himalayas is rather Harebell-like and the most suitable one for a small rockery.

CREPIS, Hawk's-Beard. There are annuals like *C. rubra* with rose or white Dandelion-like flowers and perennials like *C. aurea* with showy orange-copper ones, a native of alpine meadows, which can be a bit invasive. '*Crepis*' means a boot but the resemblance escapes me and it is not obvious that hawks have beards though perhaps the resemblance lies in the ring of fine hairs with which the fruit of the plant is surrounded.

CYNANTHUS. They are related to the Bellflowers and the two species I have seen have blue or violet Periwinkle-like flowers, so I think the name is more likely to mean 'blue-flower' from Greek *kyanos* than 'dog-flower' from *kyon*: none of my authorities sheds any light on the matter.

DODECATHEON, Shooting Star, American Cowslip. They are like slender upright Cyclamen needing moist soil and partial shade. Pliny used the name meaning 'twelve gods' for a spring flower - perhaps the Primrose - which was supposed to be under the special protection of the principal deities.

DRABA. The name means 'acrid' in Greek. It is a big genus

containing many weeds but for the rockery there are several alpines with yellow flowers. One native species, *D. verna* (of spring) was used as a poulticing plant and gave rise to the English names Whitlow Grass or Nailwort.

DRYAS octopetala. The white flower with golden stamens reminds me of an eight-petalled Blackberry flower and the dark green leaves, white beneath, are oak-like. Dryas was a wood nymph to whom the Oak was sacred. It is easier to find in the Alps than on Scottish or Irish mountains where it is a rare native.

EDRIANTHUS. They are sun-loving blue and purple Bellflowers from Dalmatia and the Balkans. Their upward-pointing bells appear to be sitting (Greek *hedraios*) on their cushion of foliage.

EPILOBIUM, Willow Herb. This is a big genus with useful European species for the larger rock- or scree-garden as well as the familiar weeds and woodland and waterside flowers. The only real dwarf is *E. kaikourense* which some authors say is Japanese but, in fact, the name comes from Kaikoura in New Zealand's South Island. The generic name means 'upon a pod', *lobos*, from the way the corolla grows at the end of the ovary.

ERINUS, Fairy Foxglove. *E. alpinus* will seed itself into any crevice and its flowers can range in shade from white through pink to deep red. The name, which was used by Dioscorides for a quite different plant and transferred to this genus, suggests early flowering, *er* being 'spring'.

ERYSIMUM. The genus now includes the Alpine Wallflowers as well as our native Treacle Mustard, *E. cheiranthoides*. The sap of some species is said to raise a blister, Greek *erysimon*, though 'Blister Plant' is a country name in some parts for the common Buttercup, *Ranunculus acris*.

GILIA, Prickly Phlox, Thimble Flower. They come from the W. coast of the U.S.A. and the dwarf hardy annuals like *G. lutea* (yellow) make a splash of colour from mid to late summer. I believe them to be named after Filippo Luigi Gilii (1756-1821) who was a Roman astronomer, but one author says 'after Gilio, a Spanish botanist', and another says 'from a Hottentot name for a plant used to make a beverage'!

GLAUCIDIUM palmatum is very handsome, comes from Japan, has flowers resembling those of the Horned Poppy, *Glaucium* (grey-green, from the colour of the foliage), and palmate leaves, that is, five-lobed like a hand.

HABERLEAS come from the Balkan and Greek mountains and like a shady crevice. They were named after Carl Constantin Haberle (1764-1832) who was Professor of Botany at Budapest. *H. ferdinandi-coburgii*, named after a king of Bulgaria, has lilac primrose-like flowers, and *H. rhodopensis* from Rhodope Mountains in the same country is similar but smaller and has a pure white form.

HEPATICA, (see p.128).

HOUSTONIA serpyllifolia (thyme-leaved) is often incorrectly called *caerulea* and known in America as Bluet. It is like a Chickweed with masses of pale blue flowers and was named after a Scottish surgeon, William Houston (1695-1733) who collected and drew flowers in Central America and the West Indies.

IONOPSIDION, Violet Cress. There is only one species, *I. acaule* (stemless), a hardy little annual from Portugal which likes moist shade. The four-petalled flowers are not really 'violet-like', the meaning of the name, except in being white or mauve.

JEFFERSONIA, Twin Leaf. They were named in honour of Thomas Jefferson who was a keen plantsman as well as being third President of the U.S.A. (1801-1809). The two species both have bilobed leaves, though the American one with white flowers is called '*diphylla*' while the lavender-flowered one from Manchuria is '*dubia*'.

LEONTOPODIUM, Edelweiss (see p.10). It is not easy to find it growing wild in the Alps; perhaps that accounts for the universal popularity of *L. alpinum* which predates *The Sound of Music*. The Greek name, meaning 'lion's foot', originally belonged to a different genus.

MAZUS, with species from the Himalayas and New Zealand, is a creeping relative of the Snapdragons. You can see the resemblance in the shape of the flowers which can be white or lilac with pretty speckling on the lip. The name comes from Greek for a teat and refers to nipple-like swellings in the throat of the flower.

MOLTKIA makes a tidier, more upright sub-shrub than its relative *Lithospermum* with which it is easily confused. The hybrid *X intermedia* is usually grown and thrives on sunshine. Count Joachim Moltke (1746-1818) was a Danish statesman and naturalist.

MORISEA. There is only one species, *monanthos*, with single four-petalled yellow flowers, which was found in Sardinia by Giuseppe Moris (1796-1869) who wrote a flora of that island and was a professor at Turin.

NIEREMBERGIA, Cup Flower. *N. repens* from Argentina is the dwarfest and hardiest of these S. American perennials with bowl-shaped, upturned white flowers. Juan Eusebio Nieremberg (1596-1658) was a Spanish Jesuit who wrote a book about the wonders of nature.

ORPHANIDESIA gaultheriodes is a shrublet from the Middle East with shell-pink flowers like a tiny Rhododendron. It manages to commemorate two botanists, one Greek and one Canadian: Theodoros Orphanides (1817-1886) was a professor at Athens and the specific name means 'like a *Gaultheria*', Creeping Wintergreen, (see p.112).

OURISIA coccinea comes from the Andes and needs moisture in summer but can be killed by too much of it in winter. Flower stems rise from a thick mat of leaves and carry drooping tubular crimson blooms. The genus was first found by Bougainville's round-the-world expedition, growing in the Falkland Islands and called after the governor whose name was Ouris.

PARAHEBE means 'close to the Hebe genus'. They are tough little evergreens from New Zealand. *P. catarractae* (of waterfalls) foams with white flowers but *P. lyallii*, discovered by David Lyall (1817-1895) is dwarfer and may be pink or purple as well.

PARNASSIA palustris, Grass of Parnassus. I cannot resist a native plant as delicately beautiful as this with its green-veined white petals. The name is from Mount Parnassus, home of the Muses, and the specific name means 'of marshes'.

PHYSOPLEXIS comosa (tufted), Devil's Claw. This is a plant of limestone alpine crevices (still sometimes called *Phyteuma*). Greek *physa* is a bladder, though to my mind the flowers look more like mauve-purple flasks, and *plexis* is 'plaited' from the way they are packed together in the flowerhead.

PINGUICULA, Butterwort, is another irresistible native if you have a patch which is wet and peaty enough. *P. grandiflora* from S.W. Ireland and the mountains of continental Europe is larger than the common *P. vulgaris*, indeed, with flowers like long-stemmed deep purple Violets and a rosette of pale green insect-eating leaves, it is one of the most beautiful and interesting of all wild flowers. Latin *pinguis* is 'fat' and refers to the greasy appearance of the leaves which were used to make a salve for cows' udders in Yorkshire in Gerard's day. Elsewhere their properties were purely magical in protecting cattle and therefore the butter from elves.

PLEIONE, Indian Mountain Orchid. Most of the species need cool greenhouse treatment but *P. formosana* (from Formosa - now Taiwan) should survive on a sheltered rockery and is as exotic looking as more tender species. Pleione in Greek mythology was wife of Atlas and mother of the seven Pleiades who were metamorphosed into stars to escape the attentions of Orion the hunter.

PRIMULA, Cowslip, Polyanthus, Auricula. Linnaeus chose the name for the genus of over 500 species, though *primula veris* (first little one of spring) had been commonly used of the Daisy as well as the Primrose (from the *prima rosa* of medieval Latin or, perhaps, since the earliest English spelling is 'Prymerolle', from the Italian *prima verola*). Together with our native Cowslip and Oxlip, it was the main ancestor of all the single and double many coloured garden Primulas and Polyanthuses, which were regarded on the Continent as peculiarly English flowers, until the introduction of the Asian species such as the Drumstick Primulas (*denticulata* = somewhat toothed) and the Candalabras such as *pulverulenta* (mealy) less than a hundred years ago. The Auricula or Dusty Miller is an alpine native cultivated since Roman times and known once as '*auricula ursi*', 'bear's ear', from the shape of the leaves.

RAMONDA. There are several similar mountain species which are hardy but dislike direct sun. They make neat rosettes of leaves from which rise stems with saucer-shaped flowers in shades of lilac and violet-blue. They are named after Baron Louis Francis Ramond

(1753-1827) who collected *R. pyrenaica* in the Pyrenees.

RAOULIA. These dwarfs make the nearest thing to short-pile, close-weave carpet in the garden. *R. glabra* (smooth) is emerald green with tiny white flowers or, if you want a grey carpet 'turning yellow' choose *R. lutescens*. They were named after a French naval surgeon, Edouard Raoul (1815-1852), who collected in New Zealand whence they came.

RHODOHYPOXIS. The two species come from S. African mountains and are not easy to make comfy here: they need well-drained, lime-free soil and moisture in the summer rather than the winter. The attractive flowers are shades of pink or red (*rhodo*) and *Hypoxis* is the name of the Star Grasses, a related genus, from Greek *hypo*, 'under', and *oxys*, 'sharp'.

SANGUINARIA, Bloodroot, gets its names from the reddish sap in the fleshy root which also produces lobed grey-green leaves and pure white flowers. The only species is *canadensis*, and the double flowered form is 'Flore Pleno'.

SANVITALIA, Creeping Zinnia. *S. procumbens* (lying flat) is a hardy annual from Mexico and well-suited to rockeries. The yellow daisy flower is similar to a *Helenium* or *Arctotis*. Frederico Sanvitali (1704-1761) was an Italian professor at Brescia.

SAXIFRAGE. Pliny called the Maidenhair Fern *saxifragum* (stone-breaker) and inferred that it would be useful for breaking up and expelling stones from the body from its ability to root in and expand cracks in rock. The name and the supposed medical use passed to the big genus of 370 species to which it now belongs. We have about a dozen natives ranging from the common Meadow Saxifrage *S. granulata* (so called from the bulbils at the base of the stem) to the very rare Clustered Alpine Saxifrage, *S. nivalis* (of the snow). The scores of good rock garden plants, many of them hybrids, may be as big as *S. longifolia* with an eighteen inch plume of flowers rising from a dinner-plate-sized rosette to tiny Moss Saxifrages, some of which are best appreciated in a pan in an alpine house. In between are vigorous spreaders like London Pride, *X urbium* (of towns).

SEDUM, (see p.5,30).

SEMPERVIVUM, Houseleek, Hens and Chickens. There are

25 species and many hybrids of these sun-loving, evergreen plants. The flowers are less valuable than the succulent foliage which gives much variety of form and colour ranging from grey and jade-green to scarlet and purple. The name means 'always alive' from their toughness and ability to survive even on the roofs of houses, which they were once thought to protect from lightning.

SHORTIAS are named after a Kentucky botanist, Dr Charles Short (1794-1863) though *S. galacifolia* (with leaves like Galax, the Oconee Bell) was actually discovered in Carolina by André Michaux and then lost for half a century. The Japanese species have prettily fringed petals reminiscent of *Soldanella*.

SILENE, Campion, Catchfly. There are typical alpine cushion plants from several parts of the world including the Moss Campion, *S. acaulis* (stemless), from our own highest mountains, as well as the familiar annuals. '*Silene*' comes from Greek *sailon*, 'saliva', because of the sticky secretion of some species which can trap insects, though the Catchflys are not actually insectivorous like Butterworts or Sundews. 'Campion' is an Elizabethan name probably originally given to one of the Mediterranean species which they thought was 'champion' (the spellings are interchangeable) when it was introduced to their gardens.

SOLDANELLA, Snowbell. They come from the high European mountains and are used to spending the winter under the snow and flowering as soon as it melts. Damp, cold and slugs are their enemies. The round leaves suggested *soldo* a small coin in Italian.

VIOLA. There are 500 species to choose from. I particularly like our native Mountain Pansy, *V. lutea*, and the Yellow Wood Violet of the Alps, *V. biflora*, which has neat tufts of kidney-shaped leaves. 'Violet' from French *violette* is obviously a little *Viola*, the Latin name. This in turn came from Greek *ion* which originally had an initial *digamma*, a letter equivalent to our 'w' or 'v' which became obsolete before the classical period.

Chapter 12

A Bushel Of Bulbs

With bulbs I include corms, tubers and rhizomes: they are all storage devices, like the camel's hump, allowing perennials to survive in difficult conditions. They may also be a means by which plants spread and reproduce themselves without the bother of sex. Some (like the camel) are equipped to survive heat and drought and may not much like the damp cold of our winters. Others (like the dormouse which stores up fat) hibernate in the cold season. It is often helpful to know where bulbs come from, whether they are camels or dormice, if you mean to treat them kindly.

The primary meaning of Greek *bulbos* is 'Onion', and an Onion which every one has dissected is a typical bulb. It consists of a stem modified into a dome-shaped disc from which grow a number of thickened storage leaves folded tightly over one another.

A corm is also a modified stem, a rounded, swollen stem-base; *kormos* actually means a tree-stump in Greek; but it is not perennial: it shrivels after one season's growth as anyone who has lifted Gladioli will know. At the top it has a bud from which the shoot and also the roots of the new corm grow.

A tuber may be a swollen underground *stem* as in the case of the Potato or a *root* as with the Dahlia and, by having eyes from which new growth emerges, they also have the dual purpose of storage and reproduction. Some tubers can be enormous: an edible yam is on record weighing 90 lbs.

A rhizome is a thickened horizontal stem on or just below the soil surface which again serves the same two functions. Corms may be produced on it, as with the Crocosmias, or not, as with most of the Irises. Perennial weeds like Couch Grass often use rhizomes for their aggressive colonization. Others, like some Oxalis species, produce masses of miniature bulbs or 'bulbils' which are scattered by cultivation and make them difficult weeds to eradicate. Bulbils may also be produced above ground as with some Lilies and Alliums.

With some of the most popular bulbs like Crocus, Gladiolus, Narcissus and Tulip there are scores or even hundreds of species, and named varieties in cultivation may run into thousands so I have generally had to limit myself to the generic names. Many books, both learned and popular, have been written about these genera and, of course, about families like the Iris and Lily.

There are very many delightful bulbs from countries much warmer than ours, particularly African ones; on one page of a gardening encyclopaedia I notice Vallota, Veltheimia and Watsonia, named after a Frenchman, a German and an English baronet; but I have had to exclude those which will not grow out of doors even in the milder parts of Britain, making an exception only for a few things like Hippeastrum which are very familiar as house plants.

ACIDANTHERA is a Gladiolus-like member of the Iris family and sometimes called *Gladiolus callianthus*. The name comes from Greek *akis*, 'a point' and *anthera*, 'anther'.

ALLIUM, the Onion family, takes its name from the Greek word for Garlic and the ornamental ones with their wide range of form and colour all have an Oniony smell which limits their popularity. They were all 'Molies' to the Elizabethans but today only a yellow-flowered one is *A. moly*. All we really know about the Moly of Homer and later writers is that it had a black root, milk-white flowers, magical properties and Tennyson found it a useful rhyme for 'melancholy'.

AMARYLLIS belladonna is a pink-flowered near-hardy relative of the Hippeastrums and often confused with *H. equestre*. 'Amarylis' is the archetypical name for a shepherdess in the pastoral poetry of Theocritus, Ovid and Virgil and *belladonna*, of course, is 'beautiful lady'.

ANEMONE. Greek *anemos* is 'wind' and 'Windflower' is a common name in English but it has been claimed that Anemone comes from Naamen, alias Adonis, from whose blood the red *A. coronaria* sprang in mythology. There are 150 species and it is the spring-flowering ones that tend to be tuberous-rooted. *A. nemorosa* meaning 'of woods' is our wild Wood Anemone and, of the familiar garden ones, *blanda* means 'pleasing', *hepatica* 'liver-like' from the leaf-shape, *pulsatilla* is 'quiverer' and *japonica* is actually a native of China.

ARUM is *aron* in Greek and *A. maculatum* (spotted) is our native Lords and Ladies. There are 15 species, *A. italicum* being grown for its attractively marked arrow-shaped leaves and contrasting yellow spathe. The so-called Arum Lilies are Zantedischias (see p.44).

BRODIAEA. There are 40 species from N. and S. America, the genus being named after the Scottish botanist James Brodie (1744-1824). There is the Californian Firecracker, *B. ida-maia* (after Ida May Burke) with scarlet tubular flowers but the species most often grown look like large blue or mauve Stars of Bethlehem.

CALOCHORTUS uniflorus from California is also known as the Mariposa Lily or Globe Tulip. The name is from Greek *kallos*, 'beautiful', and *chortos*, 'grass'.

CAMASSIA. North American Indians ate the bulb of the 'Quamash', one of the five species, and that is where the generic name comes from. The great David Douglas (see Ch.4) discovered these handsome blue, purple or white Lilies.

CARDIOCRINUM giganteum from the Himalayas can be 10 feet tall and is distinguished from a true Lily (*krinon* in Greek) by having heart-shaped (*kardio*) leaves. A Japanese relative is shorter and its Latin name *cordatum* has the same significance.

CHIONODOXA. Greek *chion*, 'snow', and *doxa*, 'glory' give us the common name. They were first found about 1842 flowering amidst the melting snow 7000 feet up a Turkish mountain.

COLCHICUM, the Autumn Crocus or Naked Ladies is a native British plant but the name comes from Colchis in the Caucasus, once famous for its poisons. The bulb is said to be pleasant-tasting which makes it all the more deadly, but it produced a valuable drug for treating gout. Theophrastus says that slaves sometimes ate it to make themselves too sick to work.

CORYDALIS belongs to the Fumitory family: *bulbosa* is purple-flowered and naturalised in Britain, while *lutea* (yellow-flowered) is a native and particularly at home on old walls. There is also a brilliant blue species from Kashmir. 'Corydalis' is the Greek name for the crested lark whose spur the flower is supposed to resemble.

CROCOSMIA. The dried flowers smell like Saffron so the name is composed of Greek *krokos* and *osme*, 'scent'. They are closely related to the Tritonias and Montbretias (see p.41) which is still the common name for the hardier garden varieties which have been produced by crossing species from S. Africa.

CROCUS. There are some 70 species, mostly spring-flowering, but the autumn-flowering ones should not be confused with Colchicum (see above) which is a Lily, whereas Crocuses are Irises. They love the sun and, unfortunately, mice and squirrels love their corms. Saffron is *krokos* in Greek.

CYCLAMEN is the Greek name and means 'circled' because of the way the stem bearing the seed capsule makes a coil. There are 15 species all from the Mediterranean area. *C. hederifolium* (ivy-leaved) was known as 'Sow-Bread' and the old apothecaries had many uses for it. Turner and Gerard both warn pregnant women against stepping over the plant for fear it will cause a miscarriage.

DAFFODIL, see Narcissus.

DAHLIA, (see p.37).

DIERAMAS come from S. Africa and the most commonly grown is *D. pulcherrimum* (very beautiful). The name is Greek and means 'funnel' from the shape of the flower, but it is the graceful arching stems that have suggested the popular names, 'Wand Flower' and 'Angel's Fishing Rod'.

DRACUNCULUS. *D. vulgaris* is also known as *Arum dracunculus*, 'little dragon'. It looks like a huge purple Lords and Ladies and smells like carrion to attract pollinating flies; more striking than beautiful.

ENDYMION, Bluebell, (see p.18).

ERANTHIS means 'spring flower' in Greek and *E. hyemalis*, 'of winter' is our Winter Aconite, welcome for its green ruff as well as its yellow cup. It is a poisonous plant; rabbits will not touch it; and that is why it was identified with the classical Aconite but it is not remotely related to the Wolf's-bane or Monk's hood.

ERYTHRONIUM comes from Greek *erythros*, 'red', but the Californian species are mainly yellow or white. *E. dens-canis*, the

Dog's Tooth Violet, is a native of sub-alpine woodlands in Europe which was introduced to England in the 16th century. Gerard supposed it 'the true Satyricon of Dioscorides', a powerful aphrodisiac, and the Elizabethans called all manner of un-violet-like things 'Violets'; even the Snowdrop was 'a bulbous Violet'. It is the pointed yellowish-white bulb which is like a dog's tooth.

EUCOMIS means 'beautiful head' and the English name for these natives of S. Africa, 'Pineapple Lily', is apt as the dense flower-spike is topped by a rosette of green leaves. *E. cosma* (beautiful) has a stem which is blotched and flowers edged with purple, and a 'pong' rather than a scent. I have grown it as a pot-plant but it and two other species, *E. bicololor* and *E. autumnalis*, are said to be 'reasonably hardy'.

FREESIA, (see p.38).

FRITILLARIA. We dealt with our native Snakes-head Fritillary, *F. meleagris*, on p.26, but there are 84 other species, about half-a-dozen of which can be established in Britain. By far the best known is the Crown Imperial whose name comes from the Latin *F. corona-imperialis*. It was brought to Europe from Turkey or Persia but is perhaps a native of the Himalayas. It was known in Shakespeare's England (see A Winter's Tale, IV.4.).

GALANTHUS nivalis, Snowdrop. Linnaeus chose the name *Galanthus*, 'Milk-Flower', and *nivalis* means 'of snow'. Although it seems so much at home in England, it is never mentioned by Chaucer or Shakespeare and was probably introduced in the 16th century. As late as the 18th century it was being called the 'Bulbous Violet' more often than the Snowdrop. Other names are Candlemas Bells, February Fair Maids and Mary's Tapers. Not much else is in flower at Candlemas, Feb. 2nd, the Feast of the Purification of the Virgin.

GALTONIA candicans, (see p.38).

GLADIOLUS like Gladdon, our native Stinking Iris, is from the Latin for a short sword because of the shape of the leaves. *G. illyricus* is (and probably always was) a scarce native growing in the New Forest and Isle of Wight but its specific name (from Illyria, i.e. W. Yugoslavia) suggests that it was first cultivated here from imported corms. In places abroad it can be a troublesome weed, hence names like Corn Flag and Corn Gladdon. *G. bizantinus* was another early introduction, from Asia Minor, but most of the 300 species come from Southern Africa and hybridization has produced the modern varieties in which scent and grace have often been sacrificed to size and vividness of colour.

HABRANTHUS, see Zephyranthes.

HIPPEASTRUMS from tropical S. America are more tender than Amaryllis Lilies which they resemble and are often called. One species, *H. equestre*, meaning 'of a horseman', is so named because the flower was thought to resemble a horse's head as seen by the rider. This seems to have suggested the generic name which has been translated as Equestrian Star, Horseman's Star or Knight's Star. Another name is Barbados Lily.

HYACINTHUS (see also p.18). Cultivated Hyacinths were not developed from our native Bluebell but from *H. orientalis* from S.E. Europe and Asia Minor. They were introduced from Constantinople in the late 16th century and called Oriental Jacynths or 'Zumbuls' from the Turkish name.

IPHEION is a name used by Theophrastus but its etymology is obscure. There ar 20 species of hardy bulbs from Central and S. America, *I. uniflorum*, the Spring Star Flower making attractive clumps in a rockery. They were previously regarded as Tritelias or Brodiaeas.

IRIS was, in mythology, the messenger of Juno and goddess of the rainbow and Irises can be almost any colour except, I think, a scarlet red. They were being cultivated in Egypt 3500 years ago and have been used for food, medicine, weaving and scent (orris-root). It is a complex genus of about 300 species divided for convenience into 'Bulbous', 'Crested', 'Bearded' (pogon) and 'Beardless' (apogon) Irises. Our native Yellow Flag is the fleur-de-lis which showed King Clovis Ist of France where to ford the Rhine with his army and so escape from the Goths. Our other native with scarlet-seeded pods, the Stinking Iris, *I. foetidissima*, has leaves which were thought to smell like roast beef when bruised so a country name is Roast Beef Plant.

IXIA means 'bird-lime' in Greek from the plant's sticky sap. These rather tender corms from S. Africa are something like Freesias in habit but with more vivid colours, dark centres and no scent.

LACHENALIA is another rather tender S. African bulb, named after a Swiss botanist, Werner de la Chenal (1735-1800). It is known as Cape Cowslip but a red-flowered species is *L. aloides*, 'aloe-like', which gives a better idea of the shape of their flowers.

LAPEIROUSIA cruenta. Semi-hardy and S. African in origin, its crimson (cruenta) flowers can be seen naturalised around the Mediterranean. It was named after a botanical Baron, Phillipe de la Peyrous (1744-1818).

LEUCOJUM, the Greek name, means 'White Violet' though the flowers are more like large Snowdrops. The Summer Snowflake, *L. aestivum*, 'of summer', is a rare native and the easiest to cultivate. The Spring Snowflake comes from central Europe and the Autumn one from the Mediterranean region.

LILIUM. Many beautiful flowers (and some girls regardless of their charms) have been called 'Lily', a name that goes back to Old English and before that to Latin and Greek *leirion*. Botanically, however, there are some 80 species in the genus which are divided for horticultural convenience into nine groups. *L. candidum* (radiant white) has been cultivated since antiquity and was, early on, associated with the Virgin Mary though the name 'Madonna Lily' is no older than the 19th century. The Turk's Cap or Martagnon Lilies (*L. chalcedonicum* = from the Bosphorus area) have also been cultivated for hundreds of years; a 'martagnon' is a Turk's cap, a type of turban. Tiger Lilies were introduced by William Kerr in the early 19th and Regal Lilies by E.H. Wilson in the early 20th century, both from the borders of China and Tibet. Most of the other species and varieties were also introduced this century.

MONTBRETIA, see Crocosmia.

MUSCARI. *M. comosum* (tufted, hairy) is the Tassel Hyacinth, a cultivated form of the common weed of pastures in S. Europe but not very commonly grown in gardens. Better known are the Grape Hyacinths and best known of the 60 species is the rapidly spreading *M. armeniacum* (from Armenia) though we have a native,

M. atlanticum, in parts of East Anglia. None is grape-purple; it is the form of the flower-spike that resembles a bunch of grapes; and a musk-scented Turkish species, *M. moschatum*, that gives the genus its name.

NARCISSUS. The Nymph Echo died of grief because the beautiful boy Narcissus could only love his own reflection and eventually pined away contemplating it, a flower springing up in his place. This is the story in Ovid which Pliny dismisses as poetic feigning. He says the name comes from the *narke* or stupor caused by the flower's narcotic scent, a belief which the Tudor herbalists shared. No culinary or medicinal uses ever seem to have been claimed for the genus, but Narcissi have featured in stories and ceremonies from the earliest times. The one referred to in Homer is assumed to be the Tasetta (little cup) or 'Bunched Narcissus' because of its strong scent and wide distribution. 'Paperwhite' and 'Cragford' are well known examples of this type today. Our own Wild Daffodil or Lent Lily is an 'Ajax' or long-trumpet type and was not originally considered a true Narcissus, hence its specific name *pseudonarcissus*. It is not the ancestor of modern Trumpet Daffodils like King Alfred: that is thought to be *N. hispanicus* (from Spain): *N. poeticus* also from Spain and Greece is the ancestor of varieties with white petals and short red cups. Jonquils are so called because they have round rather than strap-shaped leaves, like a rush which is *junquillo* in Spanish. No one has satisfactorily explained why 'Asphodel' became first 'Affodil' and then 'Daffodil' or 'Daffadowndilly' in English: in his *Names of Herbes*, 1548, Turner has both spellings with and without initial 'D;. One might have expected 'an Affodil' to have become 'a Naffodil'. Anyway, the flower replaced the Leek as a more suitable emblem at the investiture of Edward as Prince of Wales in 1911.

NERINE sarniensis, Guernsey Lily. Here we have a fancifully embroidered story and considerable geographical confusion. A ship bringing bulbs of these Lilies from Japan was wrecked on the Channel Isles and they established themselves on the sandy shores of Guernsey. '*Sarniensis*' is from Sarnia, the old name for Guernsey, and '*Nerine*' from the sea nymphs or Nereides who rescued the bulbs from the waves. It is true that they have been long naturalised in Guernsey: with the exception of *N. bowenii*, none of the 30 species of Nerine would survive out of doors elsewhere in Britain: but they are natives of S. Africa and not Japan.

NOMOCHARIS. There are 16 species from China and the Himalayas with lilac, pink and purplish saucer-shaped flowers mostly spotted towards the centre. The name is made up of Greek *nomos*, 'meadow' and *charis*, 'grace',

ORCHIS. Globally there are some 750 genera and about 20,000 species in the Orchidaceae family, roughly half terrestrial and half epiphytic. Some terrestrial ones like our Early Purple Orchid, *O. mascula* (male) grow from twin tubers like testicles and that is the literal meaning of 'orchis' in Greek. Although we have some 40 native species, none make reliable garden plants and they should certainly not be dug up from the wild.

ORNITHOGALUM. Star of Bethlehem, *O. umbellatum*, and Bath Asparagus, *O. pyrenacium* (from the Pyrenees), can be found growing apparently wild in England but the Tudor herbalists called them foreign flowers, so perhaps they are only denizens here. There are more than 100 species, some with bulbs said to be sweeter roasted than a Chestnut, but few are in cultivation. *O. nutans* (drooping) is hardy and handsome while *O. thyrsoides* (panicled) is the Chincherinchee from S. Africa. Why Ornithogalum should be so called is not obvious: the name was used by Dioscorides and means 'Bird's milk'.

OXALIS. There are 800 relatives of our native Wood Sorrel, *O. acetosella* (a little acid), but only a few are bulbous. The pink form of *O. enneaphylla* (nine-leaved), from the Falkland Islands, is one of the best for the rockery. 'Oxalis' comes from the Greek *oxys*, 'acid'.

PANCRATIUM meaning 'all powerful' is a name used by Dioscorides. *P. maritimum* is the half-hardy scented Sea Lily from the Mediterranean and there is *P. illyricum* from Illyria, the sea coast of Yugoslavia.

PLEIONE, (see p.124).

PUSCHKINIA scilloides (like a Scilla) from Asia Minor is

also closely related to Chionodoxa. It is named after Count Apollos Mussin-Puschkin who collected plants in the Caucasus early last century.

PULSATILLA, see Anemone.

RANUNCULUS is a genus of 400 species, not all of them bulbous, including our native Buttercups, Crowsfoot, Celandine and Marsh Marigold. Many of them are aquatic or like damp places and that is how they get their name which means 'little frog' (*rana* in Latin). The first corms of *R. asiaticus* were imported from Turkey in the 16th century and two centuries later there were several hundred named varieties available to gardeners, such was their popularity.

ROMULEA resembles a Crocus and *R. bulbocodium* which is most often grown is rosy-mauve like the autumn Colchicum. They come from S. Europe and are named after Romulus, the mythical founder of Rome.

SCHIZOSTYLIS gets its name from its deeply divided style. *S. coccinea* (scarlet) from S. Africa is sometimes called a Kaffir Lily, a name it shares with Clivia (see p.37).

SCILLA is the ancient Greek name for the Sea Squill, *Urginea maritima*, and may be linked to one or other or two verbs meaning either 'to dry' (because it was supposed to cure damp humours) or 'to injure' (because the bulb was reputedly poisonous). There are many species, some half dozen of which are grown as rockery bulbs. *S. sibirica* (from Siberia) is the best known; Parkinson called them 'Starry Jacinths'. *S. peruviana* from S. America is sometimes known as the Cuban Lily.

SISYRINCHIUM was Theophrastus' name for an Iris and in Greek it means 'swine's snout' because pigs grub up the rhizomes. The monk in Browning's *Soliloquy of the Spanish Cloister*, parodying Brother Lawrence's mealtime conversation, asks "What's the Greek name for Swine's Snout?" There are 100 species from the Americas of which *S. striatum* with its spikes of creamy flowers is hardy and probably the best known.

SNOWFLAKE, see Leucojum.

SPARAXIS is a S. African genus and *S. tricolour*, the Harlequin Flower, will only be happy in a warm and sunny situation. In Greek *sparasso* means to tear and the name refers to the tattered appearance of the flower's spathe.

SPREKELIA. Von Sprekelsen (1691-1764) was a Hamburg lawyer and keen gardener, who sent this 'Jacobean Lily' to Linnaeus.

STAR OF BETHLEHEM, see Ornithogalum.

STERNBERGIA, (see p.43).

TECOPHILAEA is a genus of two species named in the 18th century after an Italian botanical artist, Tecophila Billoti. *T. cynocrocus* (blue crocus) is known as the Chilean Crocus and its blue has a Gentian-like intensity.

TIGRIDIA from Mexico is spotted like a leopard yet the name from Latin *tigris* suggests 'tiger-flower'. The reason is that the spotted jaguar is known as the S. American tiger. The ones commonly grown are varieties of *T. pavonia* named after a Spanish botanist who worked in Peru, Don José Pavón (1790-1844).

TRITONIA comes from C. and S. Africa and the numerous species are related to Freesias and Crocosmias. Triton in Greek mythology was a sea god, son of Poseidon (Neptune) and depicted as carrying a conch shell, which he blew to control the waves. It is also the name of a river and lake in Libya. However, two authorities say the name means 'weathercock' because the flower's stamens point in different directions, but I can find no other confirmation of this meaning.

TULIP. It is thought that Tulips are not native to Europe - they are never mentioned by classical writers - though some species, like the yellow *T. sylvestris*, have been naturalised in England and elsewhere for a long time. It is possible the Crusaders brought them back from the Middle East but it is certain that they were being cultivated in Turkey in the 16th century and that the Austrian ambassador there introduced them to Vienna in 1554. They soon became all the rage, fortunes were spent and made and varieties proliferated. The R.H.S. now recognises 15 divisions, ranging from three foot giants whose goblets 'will hold an English pint of wine', as one old writer puts it, to four inch dwarfs for the rockery. The streaked, marbled and feathered petals, once so popular, were caused by a debilitating virus disease. In Persian and Turkish the flowers were called *Lalé* but their first importer called them 'Tulipan', perhaps by mistake, from Turkish *tul(i)band*, a turban.

VALLOTA speciosa comes from S. Africa and looks like a scarlet Clivia. It will only do out doors in the far S.W. and Channel Isles. None of the authorities I have consulted can tell me how it came to be called the Scarborough Lily. It was named after Pierre Vallot (1594-1671), a French botanist and author.

WINTER ACONITE, see Eranthis.

ZANTEDESCHIA, Arum Lily, (see p.35,44).

ZEPHYRANTHES are called after Zephyros, the West Wind and come from the western hemisphere. *Z. candida* from the Argentine is like a candy-white stalked Crocus and will multiply in a sunny border. Other species are more tender. I have grown the rosy-mauve *Z. robusta* as a pot plant for years but find it is now called *Habranthus*, 'graceful flower', which it certainly is.

Chapter 13
A Wheelbarrow Full Of Weeds

It is easy enough to say where the word 'weed' comes from: it is an Old English word from Old High German *wiota* which meant fern or bracken, historically one of the worst of weeds for man and his beasts. It is somewhat less easy to define a weed. Dr Johnson's Dictionary says, 'An herb noxious or useless' which leaves too many questions unanswered. For instance we grow and value many plants which would be noxious if eaten and have no use other than ornamental such as Monkshood (Aconitum) or the Thornapples (Datura). One book I consulted suggested, 'Any plant which interferes with the cultivation of other plants', and that covers a good deal but leaves out bothersome things that poison animals like Ragwort, or block waterways like Water Hyacinth, or colonise railway sidings, demolition sites and the gravel of my drive which all have an unmistakeable weed-like character. They are opportunists, free-booters, squatters, ever ready to take over the earth as soon as our backs are turned. They tend to belong to a fairly small number of botanically advanced families while other families seem not to produce any rogues: it is hard to imagine any member of the orchid, gentian, primula, saxifrage or violet families becoming pests - in Britain at any rate.

The adjective 'weedy' is an oddity: how did it come to mean 'weakly, lanky, anaemic' when applied to people? It is rather toughness and resilience that give weeds their status: that and their success at reproducing themselves in one way or another. There are other paradoxes: a plant may be tiresomely persistent or invasive and therefore a weed in one type of soil or situation but no problem elsewhere. This is exceptional, however, and generally their success as weeds depends on unfussiness, an ability to make themselves at home in most places: they are born carpet-baggers. Again, a weed which can be a serious problem for the commercial grower may be little trouble to the gardener. For example, Black Nightshade with its pea-sized poisonous fruits can make a crop of peas grown for canning or freezing unsaleable.

Some of our weeds like docks, Plantains and Knotgrass are known to have been here since the end of the last Ice Age ten thousand years ago and are unquestionably natives. Since then, each invasion of hunters or farmers or conquerors has brought the seeds of new species with them accidentally on their clothes or the coats of their animals or in their bags of seed-corn. An authority on weeds, Sir Edward Salisbury, once raised 300 plants from the detrius taken from the turn-ups of his trousers. Any list of weeds today will, therefore, contain a good number of aliens. They may have been well behaved at home, but become pests when introduced accidentally or on purpose into another country. The most famous example is the Prickly Pear which was taken to Australia as a drought-resistant fodder plant but soon started taking over Queensland at the rate of a million acres a year until an insect called *cactoblastis* was also introduced to put a stop to it. Anyone who has visited the Mediterranean will have seen the beautiful sulphur-yellow Cowslip-flowered Oxalis which has spread all round its shores. It is a native of South Africa but now also carpets the citrus orchards of Australia and is the worst weed in the bulb fields of the Scilly Isles, a typical cosmopolitan. Like all weeds, it likes a disturbed soil and that generally means a man-made habitat. As soon as the farmer ploughs or the gardener digs, weeds seize their opportunity. Sir Anthony Fitzherbert writing the first book of husbandry in English in 1523 mentions Thistle, Nettle and Charlock as noisome. Perhaps the simplest definition of a weed is 'any plant growing where it is not wanted'. Its advantage is that wanting is subjective and, for sure, one man's weed is another's delight. I want to be delighted again by those cornfield flowers of my childhood which the farmer has sprayed into rarity or extinction.

Old Thomas Tusser writing his *Five Hundred Pointes of Good Husbandrie* half a century after Fitzherbert is eloquent about the evils of Boddle, Cockle and Bottle and the importance of hiring day-labourers to hand-weed the corn. His 'Boddle' is our Corn Marigold, which is seldom seen today but can be so abundant that it smothers the corn completely, reflecting back sunlight even more brilliantly than Buttercups. You could once be fined for allowing it to grow in your fields. Corn Cockle, relative of the Campions, which is becoming rare, was a pest because its seeds could not be winnowed from the grain and spoilt bread and malt. 'Bottle' is the Blue Cornflower; it was cursed by the reaper for quickly taking the edge off his sickle. Tusser has nothing to say, good or bad, about the Red Poppy. My first garden had originally been a cornfield and deep digging still brought forth a crop of those cornfield weeds. However, we must now turn to troublesome weeds as most gardeners understand them. 'Know thy enemy' is prudent, if not precisely Christian advice, and if you know his names too, he is more manageable than a nameless threat; you can curse him in good round terms. But, if you are fair-minded, I believe you will be forced to admit that he nearly always has some redeeming features too.

ANNUAL AGGRAVATIONS

Before you rush to tell me, I do know that 'aggravation' comes Latin *gravis* and strictly means 'heavy, serious, grave' but I am allowing myself to use it in the colloquial sense of annoying or vexatious. Most annual weeds are just that, rather than grave threats to our horticultural enterprises like some perennials. They are easily eradicated by hand-weeding because the root (*radix*) is neither deep nor strong. Often they are vexatious because of the quantity of seed they produce and their efficiency in dispersing it. However, some do not produce an exceptional number of seeds per plant but rather have the ability to beget three or even more generations of offspring in a single season. This is true of our first three, Groundsel, Chickweed and Shepherd's Purse. All three can be found in flower in any month of the year when the weather is open.

GROUNDSEL (*Senecio vulgaris*). The genus *Senecio* contains useful shrubs and succulents, as well as annuals, particularly ones like *greyi* with silver-grey foliage, but it is the white-haired seed heads that give it its name meaning 'old man' (whence 'senile'). It has been estimated that a single plant of our common weed could be responsible for three generations and 1000 million offspring in one season, and they are at home in any soil. It is not surprising, therefore, that the name 'Groundsel' seems to come from Old English and means 'ground-swallower'. However, earlier still it may have been '*grunde*-swallower', grunde being a gungy discharge from the eyes: it was a poulticing herb for many centuries and Culpeper is still recommending it for rheumy eyes in the seventeenth century. Birds like the seeds, so it has also acquired names like Canary Seed and Chickenweed.

CHICKWEED (*Stellaria media*). Chickweeds are also liked by birds, as the name implies, and this is the commonest of them. It is the prettiest of them, however, the Stitchworts with their star-like flowers which give the genus its Latin name. Actually the weed's flowers are beautiful too, if you take the trouble to look at them with a lens; they have splendid purple stamens. The mat of growth the plant makes looks green and succulent even in the severest winter weather. No doubt poultry really do like it for it has similar names in French and German and Gerard recommends giving it to cage-birds 'when they loathe their meate'. Variations on the name include Chick-Wittles and Cluckenweed.

SHEPHERD'S PURSE (*Capsella bursa-pastoris*). It wasn't only shepherds who carried a little leather bag which looks roughly triangular when suspended from the waist by a draw-string: you can see them in Italian Renaissance frescoes, in Flemish peasant scenes, on medieval monumental brasses and elsewhere. The flattened seeds, which birds like and help to spread, could be fancied gold coins. The usual English name is a translation of the Latin specific one, *Capsella* meaning 'little box'. There are many variations on the pouch or wallet resemblance such as Pickpocket, Poor Man's Purse, Caseweed, Money-bags, Shepherd's Scrip, but other similarities are suggested by Old Woman's Bonnet, Devil's Oatmeal and Mother's Heart (because the purse breaks in half to release the seeds, and countryfolk are unsentimental about the rewards of bringing children into the world).

ANNUAL MEADOW GRASS (*poa annua*). A posh lawn should be composed of perennial grasses only but this is the commonest of the annuals which invade and its seeds get carried round the garden on our boots and with the lawn mowings. Given a chance it can flower all year round and make a lawn where you don't want one. It is not a conspicuous plant and does not seem to have attracted any country names. *Poa* is from the Greek for grass or pasturage and, in English, 'grass', 'graze' and 'green' all stem from the same ancient German root meaning to grow.

HAIRY BITTER-CRESS (*Cardamine hirsuta*). *Cardamine* was the name given to cress by that old Greek medical man Dioscorides who had served in Nero's army. 'Cress' comes from an ancient Germanic verb to nibble or eat. The most beautiful of our native Cardamines is the lilac Cuckoo Flower or Lady's Smock, but Watercress is a Nasturtium or 'nose-twister' and belongs to another branch of the large Crucifer family which includes Mustard, Cabbage, Radish, Wallflower and Stock. I am sure Hairy Bitter-Cress is becoming a commoner nuisance. It is a ballistics expert and, at a touch, will fire its seeds in all directions to a distance of nearly a yard. When it is small it tends to lurk unnoticed and gets imported by the unwary buyer with container-grown plants from the garden centre.

FOOL'S PARSLEY or DOG'S PARSLEY (*Aethusa cynapium*). Fool's Parsley's dark green deeply divided leaves do look more than a little like those of the flat-leaved variety of the culinary herb which is often grown today for its better flavour so watch out. It doesn't poison many people because it has a very disagreeable smell when bruised reminding me of mouse-droppings. It is best not to let it get to the stage of flowering in your garden but if it does, it is distinguished from other umbellates by its rather long downward-pointing bracts. *Aethusa* is from Greek for burning; 'pungent' would be a better adjective. *Cyno-* means 'dog' in many plant names, usually in the derogatory sense of 'only fit for a dog', and *Cynapine* is the name of one of the toxic alkaloids in Fool's Parsley.

ANNUAL NETTLE (*Urtica urens*). We have two native nettles that sting: everyone knows the Stinging Nettle proper, a tall perennial with tough yellow roots. There should be a place for it in a wild garden as the food plant of Red Admiral, Peacock and Small Tortoiseshell butterfly caterpillars. It was once valued as a pot- and medicinal herb and its stem-fibres have been twisted or spun and woven into cloth since prehistoric times. 'Nettle' like 'net' comes from an ancient verb meaning to twist. The Annual Nettle has equally effective microscopic hypodermic needles and is the more likely to bite the hand that tries to weed it. Both halves of the scientific name refer to stinging. Nettles love nitrogen-rich soil and if you use well-rotted farmyard manure you will almost certainly import nettle seed with it from the midden. There are not many other names: Tanging Nettle in Yorkshire, Hidgy-Pidgy or Hoky-Poky in the west country and Devil's Leaf in several areas. Did you know that 'urtication' means to beat the skin with nettles? Apparently some Romans did it to stimulate the sexual appetite!

PEARLWORT (*Sagina procumbens*). Pearlwort like the related Spurry (Spergula arvensis) is often an indicator of an acid soil in need of lime. Spurry used to be called *Sagina* too, which means 'fodder' and Spurry (but not Pearlwort) was once regarded as useful winter food for beasts. It is also attractive to poultry. Pearlwort is the more ground-hugging (procumbens), mat-forming of the two and therefore difficult to hand-weed. Hoeing merely spreads it around. It doesn't mind being trampled on and so often colonises paths and tracks. Its dust-like seed can be blown anywhere by the wind and it has even been found growing on church steeples. John Ray seems to have been the first to call it Pearlwort and both the unopened flowers and the seed capsules can look like seed-pearls. It has been called Poverty, Little Chickweed and, like many other plants, Bird's Eye.

There are plenty of other native annuals which may crop up uninvited in the garden but it will depend on where you live who these trespassers are. In my own garden there are Red Dead Nettle (Lamium purpurium), Petty Spurge (Euphorbia peplus), Knotgrass (Polygonum aviculare = eaten by small birds), and Fumitory from Old French *fumeterre* 'smoke of the earth', (*Fumaria officinalis*). I regard them with tolerance because they are attractive when not in the way and easily pulled out when they are.

Some of the perennial weeds produce a lot of seed: Docks, Dandelions, Thistles and Ragwort for example. If you have an Ash or Sycamore in or near your garden, their seedlings soon anchor themselves with a tough taproot and are not easy to tug out. Many of them, however, are weeds because of their ability to spread without the need for sexual reproduction and seldom produce viable seed. Usually they creep sideways either above or below ground but they have several other strategies, as we shall see.

GROUND ELDER (*Aegopodium podagraria*). The names Ground Elder, Ground Ash and Dog Elder must all refer to the leaf shape but so presumably does *Aegopodium* meaning goat's foot. *Podagraria*, meaning 'of gout', points to other English names like Goutweed and Acheweed,while Jack-jump-about indicates its invasive character. It was probably introduced to medieval herb-gardens but was already being cursed by Gerard in the sixteenth century for 'getting every yeere more ground to the annoying of better herbes'. The pounded leaves were laid on the gouty joint to reduce swelling and Geoffrey Grigson says it 'makes spicy and tolerable eating if the leaves are boiled like spinach'.

COUCH-GRASS (*Agropyron repens*). There is no end to the rude names this pest has been called. Even with modern chemical warfare its eradication is expensive and time-consuming. *Repens* is creeping and *Agropyron* means field-wheat. Its English names including Twitch, Squitch and Scutch all probably come from Old English *cwice* (quick) meaning living or lively. Its sharp pointed underground stems spread rapidly and will go straight through a daffodil or tulip bulb that gets in the way. Any bit left in the soil which has a joint will quickly grow again.

GREATER BINDWEED (*Calystegia sepium*). *Calystegia* refers to the fact that the calyx of the flower is at first hidden and *sepium* means 'growing in hedges' where Bindweed is often beautiful and not necessarily a menace. Some of its numerous English names refer to the white funnel-shaped flowers, e.g. Hedge Lily, Old Man's

Nightcap, Lady's Shimmy and Bellbind: rather more reflect its propensity for garrotting everything in its path, e.g. Strangleweed, Hellweed, Devil's Garter, Bearbine, Ropewind and Withywind. It climbs anti-clockwise and can make a complete revolution in little more than an hour. It can regenerate from a mere fragment left in the soil and so scarcely needs seeds and doesn't often produce any. However, when it does, they are glossy black and quite large and have been found still viable after forty years.

HORSETAIL (*Equisetum arvense*). 'Horsetail', which is descriptive enough is a direct translation from Pliny's *Equisetum* and *arvense* means 'of the field'. It is the commonest of the nine British species which are mostly bog-plants. I don't know any other names for them but I guess there must be some. They are much more primitive than the flowering plants and their tree-sized ancestors dominated the forests which became the coal measures. The upright stem, beautiful when it first pushes through in spring, is crowned by a sort of cone which produces clouds of spores. However, it is the ramifying roots which are the problem. They have small, easily detached tubers on them from which new plants can grow and on light soils and rubbish dumps the roots have been found an astonishing sixty feet below the surface. The plant contains flinty crystals which make it abrasive so that it has been used for scouring pots and even in place of fine sand-paper by cabinet-makers.

CREEPING THISTLE (*Cirsium arvense*). 'Cut thistles in May and they'll grow in a day; cut them in June and that is too soon; cut them in July and then they'll die'. Several of the family can be a nuisance to the farmer because, as this bit of country lore suggests, they can spring up again from a deep tap-root as well as producing clouds of parachuted seeds. Creeping Thistle is the worst because it also puts out horizontal roots and a single plant can soon become a sizeable thicket. *Cirsium* is the ancient Greek name for the thistle and 'thistle' itself comes via Old English from an ancient Germanic verb meaning to prick. The only other English name I know for Creeping Thistle is Scented Thistle and indeed on a still, warm day their delicious honey-like fragrance can be quite heady.

DANDELION (*Taraxacum officinale*). As with the Docks

and Thistles, Dandelions can be difficult to eradicate because of their tap-roots. One plant can also produce two thousand seeds in a season which get blown everywhere. If they get established in a lawn, the basal rosette of leaves can kill a patch of grass as big as a dessert plate. Geoffrey Grigson says "there is no satisfactory explanation" for the name 'Dandelion' which I find puzzling. Admittedly the leaves are very variable in shape but some of the mature ones with their backward-curving tooth-like segments are highly suggestive of a big cat's jaw, the 'dent-de-lion' of the French. *Taraxacum* means 'disturber' and comes from the Persian for a bitter herb. *Officinalis*, meaning 'of shops' indicates that you could get it from an apothecary as a standard medicinal herb. Variations on the name Piss-a-Bed (*Pissenlit* in French) can be found all over Britain indicating its diuretic property. There are many other English names: variations on Tell-Time and Farmer's Clock from the seed head or pappus; variations on Devil's Milkpail and Bum-Pipe from the hollow stem with its milky sap; Priest's-Crown and Monk's-Head from the bald top of the stem after the seeds have been dispersed and Dog-Posy, Horse-Gowan, Dindle and many more for no good reason that I can think of.

DAISY (*Bellis perennis*). This is 'day's eye', of course and is a weed to green-keepers and other groomers of fine turf but has always been popular with everyone else. Cultivated varieties like 'Pomponette' and 'Monstrosa' can spread into adjacent lawns. The scientific name if from Latin *bellus*, 'pretty' and other names include Billy Button, Miss Modesty and, in the North, May Gowan. Culpeper thought an infusion of the leaves boiled in asses milk effectual in consumptions of the lungs.

LESSER CELANDINE (*Ficaria verna* or *Ranunculus ficaria*). Celandine together with Crow Garlic or Wild Onion and some of the pinkish-purple flowered relatives of our native Oxalis (Wood Sorrel) can all be persistent weeds because of a characteristic not shown by the perennials mentioned so far. They all produce a mass of tiny bulbils and any attempt to get rid of them by cultivation only spreads and multiplies the problem. 'Celandine' comes by way of French and Latin from Greek for Swallow and is also called

Swallow-wort because it flowers when the Swallow arrives. There are many other names like Butterchops, Frogfoot and Foxwort but one of the commonest, Pilewort, explains the Latin name. *Ficaria* means 'little fig', suggested by the root tubers which in turn resemble piles, so that *ficus* was also used as the name for that embarrassing affliction. Celandine juice mixed with wine was used, hopefully, to bathe those and other swellings. Culpeper swore by it; it had even cured his daughter of the King's Evil he believed.

CREEPING BUTTERCUP (*Ranunculus repens*). This plant has invented yet another way of being a pest. It produces stolons or runners (hence *repens*) which root at each node like a strawberry plant, and quickly anchor themselves down with a bunch of deep white roots. One plant can cover an area of forty square feet in a season and even close mowing will not stop them spreading in a lawn. Culpeper calls the Buttercups 'fierce biting herbs' because they contain an acrid poison which becomes harmless when they are made into hay. *Ranunculus* means 'little frog' because some but not all the family are happy in damp places. 'Buttercup' is obvious enough but is not an old name. It was formerly known as Crowfoot, Giltycup and many other local names. The creeping variety has been called Creeping-Crazy, Meg-Many-Feet and Toad-Tether.

BEAUTIFUL BOUNDERS

Here are some deliberately introduced plants which may all be mixed blessings and, come to think of it, they are not all beautiful either. We have nothing quite as rebarbative as the Prickly Pear but Horse Radish, however indispensable with roast beef, is an ugly brute and reprehensible bounder. I once observed it through the hedge steadily taking over a neighbour's grass tennis court. Mostly, however, they have been brought in because they are ornamental and then found altogether too exuberant to live with in comfort. The pinkish-purple flowers of *Rhododendron ponticum* (from the Pontus area of the Black Sea coast) with their freckled throats are ornamental all right, but out of place in an English woodland and liable to make a forester see red. They are no problem for the average gardener, but some of the following may well be.

JAPANESE KNOTWEED (*Polygonum cuspidatum*). Foreign relatives of our own Knotgrass need watching or they will take over: 'Solly, this my garden now': think of Russian Vine (P. baldschuanicum)! Japanese Knotweed was introduced in the last century and described in *The Garden* in 1879 as 'one of the finest herbaceous plants in cultivation' and indeed its plumes of creamy white flowers borne head-high on stems like blotchy sugar-cane can be a brave sight. I have heard an old keeper call it Pheasant Weed, so it was probably also planted in game-coverts. The problem with it is that it spreads rapidly by means of its tough woody rootstock and getting rid of it requires much energy, a mattock and preferably a stick of dynamite. It must have been called 'cussed' more often than 'cuspid'. It is the heart-shaped leaves, tapering abruptly to a point, that are 'cuspidate'. There is rumoured to be an even larger Polygonum now on the loose from the island of Sachalin after which it is named, so watch out. Polygonum is literally 'many-knees' from the swollen stem joints which are a feature of all the family.

CREEPING SPEEDWELL (*Veronica filiformis*).From the very large to the very small. Our native blue-eyed Germander Speedwell is a plant of the wayside that speeds you on your journey. It has a host of country names. There are many less attractive little Veronicas as well, all named for some reason after the saint who wiped Christ's brow. *V. filiformis* is a foreigner. It comes from Asia Minor and the Caucasus but it is not common in those regions. It is neat and has a profusion of sky-blue flowers with a violet tinge and so it was introduced as a rock-plant a hundred years ago. It seems to have taken some time to get acclimatised but since the last war it has escaped and spread in a big way. It is now a nuisance in many parts of Europe. *Filformis* means 'thread-like' and its creeping thread-like stems enable it to roll along carpeting everything, beds, borders, paths and lawns making them look more like water than grass for a few weeks in summer. Even close mowing will not eliminate it and all the Veronicas are somewhat resistant to selective weedkillers.

SNOWBERRY (*Symphoricarpus rivularis*). I grew up calling this tough shrub 'Candleberry' but find that few people know it by that name. It is descriptive of the white waxy-looking fruits which often last all winter. 'Snowberry' seems to be the common English name though there is a popular pink-berried form and I have seen more pink candles than pink snow. It was introduced from the Americas in 1817 and you can see dense thickets of it marking the site of long-abandoned Victorian shrubberies and warning you that it may not be an advisable subject for a small garden. It spreads rapidly by suckering and will survive under the shade of mature trees though produce few berries in such a situation. *Symphoricarpus* means 'clustered-fruit' and *rivularis* 'by the stream-side' though it is also listed by nurserymen as *albus* 'white'.

SOAPWORT (*Saponaria officinalis*). By comparison with our dainty native pinks to which it is nearly related, this is a coarse creature. It was probably introduced in the Middle Ages for use by fullers in washing cloth, though Turner says he never saw it growing in England. Names like Scourwort, Crowsoap, Foam-Dock and Fullers' Grass all point in this direction and its boiled leaves do produce a lather of sorts. It looks well enough growing by a stream or running along a wet ditch but the double-flowered form, sometimes called Bouncing Betty, is grown ornamentally - a big mistake in my experience, at least on a heavy soil. I was given it once by a neighbour. It spreads itself aggressively in all directions, particularly downwards, by an incredible tangle of stoloniferous roots. Culpeper claims that it is 'an absolute cure for French-Pox' but, even if you are poxy, I would think twice about giving it garden room.

CYPRESS-LEAVED SPURGE (*Euphorbia cyparissias*). There are about 2000 Euphorbias including our Christmas Poinsettia and many handsome perennials. All of them have a milky sap, irritating to some skins, which was used of old medicinally by Euphorbus, physician to King Juba of Mauritania. Our natives, like Petty Spurge and Sun Spurge, can be weeds but are seldom a serious nuisance. Of the ones planted ornamentally, three can easily get out of hand.

The splendid Caper Spurge (E. lathyrus) has established itself in some woods on the chalk in southern England and will pop up in old gardens where it has not been seen for many years. It was used as a powerful purgative, hence the name 'Spurge' by way of French from Latin *expurgare*. The seeds should not be used as substitute capers, which they resemble: eaten in quantity they can be fatal.

E. robbiae, named after Mary Anne Robb who introduced it from Turkey, spreads rapidly vegetatively and tolerates shade. Its lime-green flower spike rises from perennial dark leathery rosettes of leaves giving a contrast of colours similar to Stinking Hellebore.

The most invasive of all the Euphorbias, however, is the diminutive *E. cyparissus* with fine yellow-green bushy leaves somewhat like a conifer's needles (hence *cyparissus* = cypress-like) which turn golden and russet in autumn. Its flowers are powerfully and not unpleasantly scented, but it very rapidly colonises even clay soils and should only be used as ground cover where it can be safely confined.

WINTER HELIOTROPE (*Petasites fragrans*). Winter Heliotrope comes from the western Mediterranean, was introduced by gardeners early last century often to their subsequent discomfort, and is now established on hedge-banks and road-sides. 'Heliotrope' in this context refers to the vanilla-like scent of the pale mauve flowers which rise up in January or even December and has nothing to do with turning towards the sun. The sight and scent of it can give midwinter cheer but it is a close relative of Butterbur and Coltsfoot: all three produce their flowers before the leaves appear and can be rampant colonisers by means of their spreading rhizomes. They all have broad leaves, too, which enable them to shade out competition and these suggested Dioscorides' name *Petasites* from Greek *petasos*, a broad-brimmed sun-hat. 'Bigge inough to keep a mans head from raine', says Gerard the damp Englishman of Butterbur, which was also used to wrap butter in his day.

'The Kitchen Garden'
from John Parkinson's *Paradisi in Sole*, 1629: 1. 'Skirrit'
2. Parsnip, 3. Carrot, 4. Turnip, 5.'Narewe', 6. Black Radish,
7. Common Radish.

Chapter 14

Variegated Vegatables

To variegate means 'to alter the appearance of, to impart variety to' and when you reflect that the Red Cabbage, the White Cauliflower, the Purple Broccoli and Spring Greens, not to mention that oddity the Brussels Sprout, are all varieties which have been bred from the wild Cabbage (Brassica oleracea) of our sea-cliffs, you will see the point of the title of this chapter. Brassica is Pliny's name. The Cabbage is one of the Crucifers, a family as important to the vegetable- as the Umbellifers are to the herb-garden: it includes Turnip, Radish, Swede, Sea Kale and Cresses not to mention Horse-radish and Mustard. Or think again of another sea-shore plant, the straggling Goosefoot *Beta vulgaris*: from it have been conjured Beetroot, Sugar Beet, Mangel Wurzel, Spinach Beet and Swiss and Ruby Chard.

Most of the other vegetables in the greengrocer's shop have been cultivated for millennia rather than mere centuries and it is often uncertain from what wild plants they originate or where they grew. They are very likely to have been given specific names like *oleraceus* or *sativus* which mean 'cultivated', or *edulis* or *esculens* which mean 'edible, good to eat'. 'Vegetable' is here used in a wide (though not the widest) sense: it includes tubers and roots, stems and shoots, leaves and flowers, seeds and fruit. There is no part of a plant that has not been found useful for food. Tomatoes are fruits obviously, but so, too, are Runner Beans, Cucumbers, Marrows and Mange-tout Peas. I shall risk the wrath of the botanist by leaving Rhubarb to another chapter devoted to fruits though it is nothing of the kind: common rather than botanical sense tells me that you would no more eat Rhubarb with gravy or a vinaigrette dressing than you would put custard on Runner Beans.

Excluded from this chapter is the exotic produce of foreign lands which is available today in splendid and unprecedented variety in the better class of shop. This is a pity since some have interesting etymologies: Avocado means 'testicle' for instance, and the obscure Aubergine is believed by some authorities to have meant 'anti-fart plant' originally. All manner of unpromising wild things like Stinging Nettles have been eaten at times of famine in the past and by students of Richard Mabey's *Food For Free* today and they, too, have been omitted. For reasons of space we must limit ourselves to those vegetables commonly grown in British gardens.

ARTICHOKE, GLOBE (*Cynara scolymus*). The *Cynareae* is a division of the huge family of composite plants and *Cardus* is the genus within it which contains most of our native thistles. It is not only the flower heads of this spectacular member of the family which are edible, but the young leaf shoots known as chards (from *cardus*) can be blanched and eaten like celery. *Scolymus* comes from the ancient Greek name for the plant while 'Artichoke' is from Arabic *al kharsuf* by way of Italian and Spanish.

ARTICHOKE, JERUSALEM (*Helianthus tuberosus*). This is not an Artichoke, but was thought to taste like one when it was first introduced; neither has it anything to do with Jerusalem. It came to Europe from S. America in the early 17th century and was recognised as a tuberous sort of Sunflower (Helianthus). In Italian Sunflower was *girasole* because it girates or turns with the sun, *sole*, and the English who never seem to have been good at languages took this for Jerusalem.

ASPARAGUS (*Asparagus officinalis*). A *plumosus* is grown for its ferny foliage but the edible *A. officinalis* has been cultivated from early times and the name is the Latin and Greek one from a root meaning 'to sprout'. It is a native of sandy shores in the south and west of England, which gives a clue to the type of soil it prefers, and growers used to dress their Asparagus beds annually with salt. The name comes to us from Old French *esparge* first as Sparage then Sparagras and finally Sparrow-grass, which Dr Johnson correctly said was 'corrupt', so we pedantically returned to the Latin name.

BEAN, BROAD (*Vicia faba*). The name 'Bean' is Old English and related to German *Bohne* and Dutch *boon*. It is an element in many English place names for beans were an essential item in the poor man's diet providing much of his protein. Like peas, they were a field crop, harvested, stacked and threshed like corn. We lose much of their food value when we eat them young and tender. If you want to be 'full of Beans', a phrase originally used of a spirited horse, use them dried. They were the *faba* of the Romans and I see that tinned Broad Beans imported from Italy are called 'Fava Beans'. No other flower in our gardens is black and white and few smell sweeter.

BEAN, FRENCH (*Phaseolus vulgaris*). So called 'French' Beans (like Scarlet Runners) in fact come from the New World and varieties have been called many things such as Kidney Beans, Snap Beans, String Beans, etc. The French, of course, call them *Haricots* or *Flageolets*, which is literally a musical instrument of the recorder family. *Phaseolus* is from the Greek for a bean.

BEAN, RUNNER (*Phaseolus coccineus*). The French call them *Haricots d'Espagne* but they come from tropical America and some of the best varieties are white- or purple-flowered rather than scarlet (coccineus). For several centuries after their introduction around 1600 they were grown purely decoratively to cover arbours and trellises.

BEETROOT (*Beta vulgaris*). The Beetroot, White Beet, Sugar Beet and Spinach Beet are all descended from a useless-looking wild plant of our sea shores, *B. maritima*, a member of the *Chenopodiaceae* meaning 'Goosefoot family'. So is the Mangel Wurzel, which is from German *Wurzel* - root, and *mangold* - beet, which suggested in English the old word 'mangel' meaning 'want': the French also call it 'root of scarcity'.

BROCCOLI (*Brassica oleracea var. botrytis*). In Italy Cauliflowers grow best in the winter so they are sown in autumn and harvested in spring and 'Broccoli' is the Italian name meaning literally 'little sprouts or shoots'. It would save much confusion if we called the curding sort 'Winter Cauliflower' and kept 'Broccoli' as the name for the hardy purple, green or white sprouting kinds. *Botrytis* means 'bunched like grapes' or 'having flowers in a raceme'.

BRUSSELS SPROUTS (*Brassica oleracea var. gemmifera*). *Gemmiferus*, you might have guessed, means 'bearing buds' and this queer Cabbage with the Walnut-sized buds in the leaf-axils of the stem probably did originate in Belgium. The French call them *chou de Bruxelles* and the Germans, *Kohlsprosse*. They were not widely known in England until the middle of the 19th century.

CABBAGE (*Brassica oleracea var. capitata*). There are some spectacular ornamental varieties for the flower border and the stem of the Jersey Cabbage is made into a walking-stick but such a basic vegetable has to find its place in this chapter. There are some 500 varieties available which have been developed by centuries of cultivation and some of them can be grown almost anywhere in the inhabited world. Theophrastus believed there was a natural antipathy between the Cabbage and the Vine and says they should be kept well apart which leads daft old Culpeper to say that Cabbage cures drunkenness. *Oleracea* means 'cultivated for food, edible' and it is not merely edible but delicious when not boiled to death. The English, who are the worst offenders, used to call the whole tribe 'Coleworts' from Old English *cawel*, 'stem or stalk' (Latin *caulis*). They were also sometimes called 'French woorts' in the Middle Ages and 'Cabbage' comes from Old French *caboche*, 'head'. The origin is obscure but one suspects some connection with Latin *caput*. A 'cabbage' is also a snippet of cloth appropriated as a perk by the tailor from a customer's material and, of course, a helpless, brain-damaged person.

CALABRESE (*Brassica oleracea var. Italica*). Calabrese, the Italian green sprouting Broccoli which is not frost-hardy, has become popular in the last few years. Some very attractive varieties like Romanesco, which are widely grown in northern Italy, have only recently been discovered by English gardeners.

CARDOON (*Cynara cardunculus*). It is not the flower-head of this relative of the Globe Artichoke which is eaten but the thickened stem and leaf midribs. It became a troublesome weed when introduced to Argentina but could safely be grown more often in this country. *Cardunculus* means 'little thistle' (though it is a handsome, tall plant) from Latin *cardus* which also gives Cardoon by way of Old French *cardon*, which had the wider sense of 'prickly-headed plant'.

CARROT (*Daucus carota var. sativa*). The wild Carrot will produce a good edible root with two or three generations of cultivation so it is surprising that it was not grown in England until the late 15th century, when it was introduced from France. It was then a purple or white root; what we think of as the essential carrot colour of orange-gold was only achieved after centuries of selection. It was widely cultivated in classical times, *Daucus* being its Latin name and 'Carrot' coming via Latin from Greek *karaton*.

CAULIFLOWER (*Brassica oleracea var. botrytis*). See also under Broccoli. The Italian is *cavolfiori* or 'flowers of cole' and this is probably what Gerard was adapting as 'Coleflorie' in his 16th century Herbal. Culpeper thought the whole family of Coleworts gave you wind: 'yea, as windy meat as can be eaten, unless you eat bag-pipes or bellows'.

CELERIAC (*Apium graveloens var. rapacium*). *Apium* is the name used by Pliny for an umbellifer and it is shared today by Celery and Parsnip. Celeriac is the Turnip-rooted Celery and the name does not seem to be older than the 18th century. It was presumably formed by analogy with words like cardiac, elegiac, aphrodisiac and so on.

CELERY (*Apium graveolens var. dulce*). Wild Celery is found in ditches and other wet places usually near the sea. John Evelyn knew it as 'sellery' in the 17th century but it does not seem to have a long history of cultivation in this country. The name seems to have arrived in English by way of French célerie (d'Italia) or Italian *sellerie* from Latin *selinum* and Greek *selinon*, so it must have been valued elsewhere down the centuries.

CHARD (*Beta vulgaris var. cicla*). This splendid, easy vegetable is known variously as Swiss Chard, Beet Chard, Sea-Kale Beet (not to be confused with Sea-Kale) and there is a handsome variety with red stems known as Ruby Chard. The broad mid-rib is sometimes cooked and served separately from the Spinach-like leaf. 'Chard' from Latin *cardus*, 'Thistle', was the name originally given to the blanched stems of the Globe Artichoke q.v.

CHICORY, SUCCORY (*Chicorium intybus*). The commoner English name comes via French and Latin from Greek *kikhorion* whereas 'Succory' is from the same root via Dutch or German. It may be a native of Egypt. *Intybus* is the name found in Virgil for Chicory or Endive. A pretty weed of waste places with its sky-blue flowers, the whole plant is very bitter-tasting. In cultivation, the blanched 'chicons' are good braised as a vegetable or crisp in a salad and there are attractive red or variegated varieties to add colour. The root is used as a substitute for or addition to coffee. Culpeper finds it good for swooning, passions of the heart, and many other conditions besides. *Witloof* is the Flemish name, meaning 'white leaf'.

CORN, MAIZE (*Zea mays*). The word 'corn' was used in Old English and related Germanic languages for cereal seeds or 'grain', which is actually a related word: remember that we talk about a pepper-corn: even more anciently the root meaning was 'worn-down (milled) grain', hence 'worn away, grown old'; so Greek *geron*, 'old man' and our words like 'geriatric' belong to the same bit of language history. *Zea* is the Greek name for another cereal.

The use of 'Corn' in American English for Indian or Sweet Corn only dates back to the 18th century. 'Maize' comes by way of Spanish from the *mahiz* of the Taino people of the Caribbean. Columbus found it growing in Cuba and brought it to Europe. It has never been found growing truly wild and may have originated as a hybrid between grasses in Mexico but its parentage is uncertain. New strains make it nowadays a more reliable garden crop in England's unreliable summers.

CORN SALAD, LAMB'S LETTUCE (*Valerianella locusta*). In France it is *mache* and it is more commonly cultivated on the continent than here, though it was known to the Elizabethan herbalists as a salet plant and is a common enough cornfield weed - hence the name. The succulent, spoon-shaped leaves grow in a rosette and have a taste something like lettuce. The Valerian family of which this is one is variously said to be from Valeria, the Roman province on the Danube, or from Latin *valere*, to be strong or healthy. *Locustus* means 'having spikelets' but you should remove the flower spikes to produce a good crop of basal leaves.

COURGETTE, see Marrow.

CRESS (*Lepidum sativum*). Watercress is a different matter, see below. There are in fact several Crucifers whose young leaves can be added to a salad but this one, not a native plant, is the most commonly grown and sometimes called Garden Cress or Pepperwort. *Lepidus* means 'neat, charming, elegant', and 'Cress' is the Old

English name with related names in other Germanic languages. It is a common element in minor place names like Cresswell and Carslake.

CUCUMBER (*Cucumis sativus*). Cucumbers come from S. Asia and belong to a genus which includes the Melons. They have been cultivated from ancient times and Pliny says that the Emperor Tiberius was particularly partial to them. The name is *concombre* in French, *pepino* in Spanish and the eructative *Gurcke* in German: Dutch *ajurkkijn*, 'little cucumber', gives us 'gherkin'. In the 17th century Culpeper said Cucumber juice would relieve sunburn and remove freckles from the face, and in the 18th the common spelling and pronunciation was 'Cowcumber'.

ENDIVE (*Chicorium endiva*). 'Endive' comes from the same root as *intybus*, the specific name of Chicory (see above): in Greek it was *entubon* and, beyond that it may be from the Egyptian name for January when, apparently, it was eaten. The two plants are closely related and often confused: Endive is the more lettuce-like and the Moss Curled and Staghorn varieties are quite distinct in appearance. Both make a similar, slightly bitter contribution to a salad and are usually eaten wholly or partially blanched.

HOP (*Humulus lupulus*). As a Kentish man by birth, I have to mention in passing that Hop shoots are eaten like Asparagus or made into Hop-top soup not only in that county, but in France where they are *jet de hublon,* and elsewhere on the continent. The root meaning of 'Hop' in the Germanic languages probably has to do with groping or encircling.

KALE (*Brassica oleracea var. acephala & B. napus*). Just as 'lang' is the northern form of 'long' and 'strang' of 'strong', so Kale is the northern form of Cole (see under Cabbage and Cauliflower above). What cooks and farmers call Kale today are various non-heading (acephalic) varieties of Cabbage and Rape. There are 'Scottish', 'Asparagus' or 'Curly' Kales and plain-leaved varieties, which are the hardiest, known as 'Cottager's', 'Thousand-headed' and 'Hungry Gap'. Sea Kale (Crambe maritima) is another Crucifer of the sea-shore with crisped and curled blue-green leaves which are usually eaten blanched.

KOHL RABI (*Brassica caulorapa*). *Caulo-rapa* is 'stem-Turnip' or 'cole-rape'. It is grown mainly as cattle fodder and tastes like Cabbage stalk to me though some people like it used, cooked or raw, much like Turnip or Radish. 'Kohlrabi' is the German name adapted from Italian *cavalo rapa*.

LAMB'S LETTUCE, see Corn Salad.

LEEK (*Allium porrum*). Pliny says that Nero ate Leeks to improve his voice: can this be a clue to all those singing Welshmen? It is not a native plant in the Principality but has been cultivated in Egypt and all over Europe from the earliest times and St David famously advised his fellow countrymen to wear it in their caps to distinguish them from their enemies. Cock-a-Leekie soup (like whisky) helps the Scots to endure their climate. *Porrum* was the Romans' name for various members of the Onion family and 'Leek' comes from Old English with related names in the other Germanic languages such as *Lauch* in German.

LETTUCE (*Lactuca sativa*). Lettuce does not occur anywhere in the wild but there are a bewildering number of varieties in cultivation. It is thought to be Asian in origin. I asked myself 'Why Cos?' and found the answer is that it is named after the Aegean island of that name. The Latin name means 'milky plant' and 'lettuce' comes from it by way of Old French. It is believed that the noble Roman family, the Latucini, took their name from the plant. Inevitably it was prescribed for nursing mothers under the old Doctrine of Signatures. Beatrix Potter's Flopsie Bunnies found it soporific and, indeed, in sufficient quantity, it is so. Culpeper additionally claimed that it 'abates bodily lust' and 'represses venerous dreams being outwardly applied to the testicles'.

MARROW (*Cucurbita pepo*). There is much confusion about the names of the Cucurbits: Pumpkin, Squashes, Gourds, Calabazas and Melons are all members of the genus as well as the Cucumber. The specific name *pepo* means 'sun-cooked', i.e. edible when ripe. Courgette and Zucchini are the French and Italian names for varieties of baby Vegetable Marrow and the primary meaning of 'Marrow', of course, is the soft stuff in the cavities of bones. Adjectivally it is used of Marrowfat Peas. The name 'Vegetable Marrow' seems to have been given to a new variety of Cucurbit when it was introduced in 1816, presumably in the sense of 'hard outside but soft and edible within'.

MUSHROOM (*Agaricus campestris*). Quite a lot of our native fungi are edible and several which are not difficult to find or identify are delicious. However, the Field Mushroom, *campestris*, and the closely related cultivated variety which is thought to be descended from the longer-stalked wild *A. brunnescens* are the only ones known in most English kitchens. 'They order this matter better in France', as Sterne says (but also have more fatalities caused by poisonous species). 'Mushroom' comes from late Latin by way of Old French *mousseron*. *Agaricus* is the old generic name for Field and other Mushrooms many of which are now included in other genera: it may originate from the town of Agaria, now in Poland.

ONION (*Allium cepa*). Shakespeare puns on the name which must have rhymed with 'union' in his day and also meant a large pearl (see the last scene of Hamlet). It comes by way of Old French *oignon* from Latin *unio*, *unionem*, because the leaves are *united* to form a bulb. The Romans were using *unionem* as an alternative name to *caepa* by the 1st century A.D. Onion juice mixed with honey and oil was once a common cough remedy.

PARSNIP (*Pastinaca sativa*). The wild Parsnip of our chalky grasslands will soon make an edible root if grown in good deep soil. It has been an important food crop since prehistoric times, though the sweetish taste is not to everyone's liking. As cattle fodder, the leaves are said to produce rich milk. The old name for an edible root was 'Neep' or 'Nep' (Latin *napus*: see also Turnip). The first part of 'Parsnip' comes through French from the Latin name *Pastinaca* which is found in Pliny and has been grafted on to the old 'root' name. The author of *Doctor Zhivago* has a surname which means 'Parsnip' in Russian.

PEA (*Pisum sativum*). 'Pea' comes from the older *pease* which was mistakenly felt to be plural. The Old English was *pise* from the Latin *Pisum*. There is a long history of cultivation and therefore many varieties of this European native. In America they tend to call the Garden Pea the 'English Pea' and to distinguish 'Shuck Peas' from 'Sugar Peas' - the Mangetout of the French, which are also known as 'Chinese' or 'Snow Peas' and feature largely in oriental cookery. Old Tusser writing his *Good Husbandrie* in the 1570s and using the old plural tells the farmer or his wife,

"Good peason and leekes, to make poredge in Lent
And peascods in July, save fish to be spent."

This 'poredge' would win the love of his labouring men and provide them with cheap protein. It must have been like the still popular mushy peas or the traditional pease pudding.

POTATO (*Solanum tuberosum*). Many whole books have been written about this indispensable native of the high Andean valleys which is related to the Tomato and to more sinister plants like Nightshade, Mandrake and Apple of Sodom. It was not very widely cultivated in Britain until the middle of the 18th century, except in Ireland where it quickly became such a staple food that more than half a million people starved when blight destroyed the crop in 1846. 'Potato' comes via Spanish *patata* from the *batata* which Columbus encountered in Haiti. This was the Sweet Potato (Ipomea patatas) which was introduced to Europe a century earlier than *S. tuberosum*, which Gerard knew as the 'Virginia Potato', an expensive rarity, compared with 'common' potatoes. *Solanum* is a name found in Pliny and means 'giver of solace'. A spud or spudder was originally the name of a digging tool used, among other things, for planting and lifting Potatoes.

PUMPKIN (*Cucurbita pepo*). In Greek *sikuos pepon* meant 'ripe Sikuos or Melon'; thus *pepon* was originally the nickname for a type of Cucubit eaten when fully ripe. As so often, the name seems to have reached English by way of Latin and Old French where it was *pépon* or *pompion*. The diminutive suffix -kin is singularly inappropriate for a fruit which can become big enough to make a coach for Cinderella. The English treat it as an inedible joke, though Pumpkin pie is a national dish across the Atlantic and the flowers are delicious even if you do not like the flesh of the fruit.

RADISH (*Raphanus sativus*). Old English *raedic* comes from Latin *radix*, 'root', which already had the additional meaning of 'radish' in Roman times. We have a wild Radish, *R. raphanistrum*, which could easily be mistaken for Mustard but for the purple lines in the flowers, but there is some doubt whether cultivated varieties are botanically the same species. They have been grown since ancient times, particularly in the Far East. Most people only know the round red ones, but there are many shapes, sizes and colours including a purple-black Winter Radish as big as a tennis ball. Indeed, on the island of Sakarujima in Japan, a radish weighing approximately 34 lbs has been bred, a spectacular product of skilful selection. The tops as well as the roots of the Radish can be eaten cooked or raw.

SALSIFY (*Tragopogon porrifolius*). *Tragopogon* is Greek for 'goat's beard' and Salsify is closely related to our native T. pratensis: *porrifolius* means 'Leek-leaved'. The origin of the Italian name *erba salsifica* is apparently unknown. In England it is often called the Vegetable Oyster because of the flavour of the Parsnip-like root or John-go-to-bed-at-noon because of the early closing of the daisy-shaped purple flowers. The blanched 'chards' can also be eaten like Asparagus.

SCALLION (*Allium fistulosum*). 'Scallion' like 'Shallot' q.v. comes from Ascalon, the port in S. Palestine. This multi-stemmed salad Onion with only a slightly enlarged base to the stem actually comes from Siberia but goes by a confusing variety of names such as Welsh Onion, Green Onion, Stone-Leek, Holster and Ciboule: *Fistulosum* means 'hollow' or 'pipe-like' and could apply equally well to the leaves of most of the Alliums.

SCORZONERA (*Scorzonera hispanica*). The herbalists' Latin *scorzonera* means 'snake-plant' because it was believed to cure snake-bite. It is a native of central and southern Europe not just Spain (hispanica). It is sometimes called Black Salsify (because of the colour of the roots) to distinguish it from 'White' Salsify q.v.: it is used as a vegetable in precisely the same way.

SEA KALE, see Kale (!)

SHALLOT (*Allium ascalonicum*). The French call them *éschalote* and Pliny knew them as *caepae* from Ascalon in Palestine. Cookery books often speak of their 'mild' flavour but, grown in my garden, they are second only to Garlic in pungency. They store exceptionally well and can be used in spring when all the Onions have shot.

SORREL (*Rumex acetosa*). It has many country names such as Sour Dock, Spinach Dock and Green Sauce. *Acetosa* means 'acid' or 'sour' and *Rumex* is the name of the Dock family which gives us sixteen native species and the alien French Sorrel, *R. scutatus*, ('little round shield' from the leaf-shape). *Surelle* is diminutive of French *sur*, 'sour'. It is a cure for scurvy and was once used by cooks much like lemon or vinegar (*vin aigre*, 'sour wine').

SPINACH (*Spinacia oleracea*). The Chinese name for Spinach means 'Persian vegetable' and there is good reason to think the Moors took it and its Persian name *aspanaj* to Spain where it became *espinaca* (see p.3). It is related to our native Goosefoots (-feet?). *Oleracea* means 'good for food', a fact affirmed by Popeye's biceps.

SPRING ONION, see Scallion.

SQUASH, see Marrow and Pumpkin.

SWEDE (*Brassica rutabaga*). The Swede was introduced from Sweden in 1781 and is botanically distinct from the Mangold (see under Beetroot) and the Turnip, though it is also known as *B. napus var. napobrassica*. *Rutabaga* is the Swedish name and also the one by which the root is known in the U.S.A. The French, who do not appreciate its culinary merits, call it *chou-navet*, 'cabbage-turnip', *navet* also having uncomplimentary overtones: to have *sang de navet*, 'turnip blood' is to be anaemic, listless, feeble, and *Des navets* is 'Not a chance! Nothing doing!'

TOMATO (*Lycopersicon esculentum/Solanum lycopersicum*). Tomatoes were being cultivated in pre-Columbian Peru and Mexico and the name comes by way of Spanish from Aztec *tumatl*, 'swelling fruit'. *Lycopersicon* is literally 'wolf-peach' giving an idea of the deep suspicion with which the fruit was regarded when it was first introduced to Europe in the 15th century, presumably because of its poisonous relatives in the Solanum family (see Potato). In England they were first called 'Love-Apples' from their supposed aphrodisiac properties and in 1723 were said to be eaten only by 'Spaniards, Portuguese and Jew families'. Dr Johnson does not include 'Tomato' in his dictionary and defines 'Love-Apple' economically, but unhelpfully, as 'a plant' suggesting lack of familiarity with it. Philip Miller in the 1760 edition of his *Gardener's Dictionary* calls them 'Love Apples' when grown ornamentally and 'Tomatoes' when raised on hot-beds 'for soups'.

TURNIP (*Brassica napa*). The first part of the name 'Turnip' is obscure but possibly has the sense of 'turned' or 'rounded', and the second element is 'root' as in Parsnip q.v. Turnips as winter feed for cattle were a crucial factor in the agricultural revolution of the 18th century; the Norfolk Viscount and innovative farmer was nicknamed Turnip Townshend. However, small two-inch ones make a delectable vegetable sprinkled with a little pepper. The strong-flavoured greens should not be sneezed at either: the Chinese give them the delightful-sounding name *mu-ching-choi-chai*.

WATERCRESS (*Nasturtium officinale/Rorippa nasturtium-aquaticum*). The Old English name was 'Well Cress' and it was already being cultivated in Kent for the London market in Regency times. Two hundred years before that Culpeper knew it as 'powerful against scurvy'. Country people have always gathered it from streams and ditches and run the risk of getting liver-fluke. Some say it has the best flavour when the leaves have a bronze tinge. *Nasturtium* is 'nose-twister' from the pungent taste and the newer name for the genus *Rorippa*, is from the old Saxon name. For the confusion with Shamrock, (see p.25).

WELSH ONION, see Scallion.

ZUCCHINI, see Marrow.

Nicholas Culpeper, 1616-54,
soldier, rogue-physician and astrologer whose herbal
has been in print for over 300 years.

Chapter 15
A Peck of Pot-Herbs

Theophrastus divided the plant world into trees, shrubs and herbs and 'herb' is still being used in that comprehensive sense in King James's Bible of 1611. Two centuries earlier Wyclif was distinguishing useful plants as 'erbis of vertue': the 'h' was not added until the fifteenth century and not pronounced until the nineteenth. By the time of the great herbalists like Gerard and Parkinson it was common for writers to distinguish between strewing-herbs, nose-herbs, dye-herbs, salet-herbs, sweet-herbs and pot-herbs. Simples or medicinal herbs were further subdivided into purging-herbs, vulneraries, emetics and so on. They were, of necessity, more 'herbivorous' than we are. 'Digestives' might be eaten both for their flavour and their real or supposed health-giving properties, there was no hard and fast division; nor were they all strictly 'herbs' in Theophrastus' sense as some like Bay, Sage and Rosemary are distinctly shrubby. The majority, however, were herbaceous in the botanical sense of non-woody plants dying completely or down to the ground after flowering.

Home-grown herbs (and imported spices) were of great importance to our forefathers not only as medicine but also in the kitchen because the basic ingredients of their diet, by modern standards, were very monotonous and limited. Today we would call many of their potherbs 'vegetables', but the word has only had that sense for about two hundred years: you will not find it in Dr Johnson's Dictionary and he speaks of potherbs as providing 'nourishment'. The 'sweet-herbs' of former times have become the potherbs of today's kitchen cupboard (and of this chapter) because they are added to the cooking or serving of a dish primarily for their flavour rather than their food value.

There has been a great revival of interest in herbs in recent years, starting perhaps with 'Flower-Power' in the 1960s and part of the larger 'green' movement. Many forests must have been pulped to produce the scores of books that are now available on every aspect of their history, cultivation and use. Few useful herbs are not also decorative, and restored or newly designed herb gardens are an attractive and interesting feature of many gardens open to the public. Perhaps the most ambitious is the National Trust's garden at Acorn Bank near Penrith where two hundred different herbs are grown in what was formerly a walled kitchen garden.

Many of the culinary herbs are natives of the Mediterranean region, were probably first introduced to this country by the Romans and will thrive best in sunny and well-drained situations. The essential oils which give them their aromatic character fortunately also seem to make them less susceptible to attack by inspect pests than most of the things we try to cultivate. Many of them have been given specific names like *odorata*, *graveolens* (strong-smelling), *sativus* (cultivated), or *officinalis* (of shops), as we shall see.

There are, of course, several distinct species of some familiar herbs like Mint and Thyme but we will limit ourselves to the one or two most commonly used in cooking. The plant family most frequently represented is the *Umbelliferae*, which also includes important vegetables like Parsnip, Carrot and Celery, but it is wise to remember that it also has some very poisonous members like Water Dropwort, Fools Parsley, Giant Hogweed and Spotted Hemlock, which can be confused with the useful, edible ones.

ANGELICA (*Angelica archangelica*). Candied Angelica is now mainly imported from Spain and the south of France, but in the London of 1665 the herb was still being used to ward off the plague. The name comes from the legend that its efficacy against epidemics like cholera was revealed by the Archangel Raphael.

ANISE (*Pimpinella anisum*). Aniseed is the seed of Anise and may have been thought to look like pepper-corns (*piper* in Latin) which is, perhaps, the explanation for the otherwise obscure medieval name Pimpinella. Anise comes through Latin from Greek *anneson*. It was valued as a digestive as well as a flavouring for food and drink.

BALM (*Melissa officinalis*). 'Balm' comes by way of Latin and Greek from Hebrew *basam* (Balsam) meaning scent. Anyone who has grown it will know that its lemon scent is attractive to bees and Melissa in Greek mythology is the name of the Nymph who was supposed to have discovered the use of honey. Balm has long been believed to promote good spirits and longevity.

BASIL (*Ocimum basilicum*). 'Basil' comes from Greek *basilikon* meaning 'princely' or 'royal'. It was supposed to have been found growing round Christ's tomb after the resurrection and it has been used in preparing holy water in some Greek Orthodox churches. Combined with garlic and tomatoes it certainly makes a princely dish. *Ocimum* comes from the Greek for an aromatic herb.

BAY (*Laurus nobilis*). 'Bay' comes by way of Old French *baie* from Latin *bacca*, 'a berry', especially the fruit of the Olive. In warmer climates than ours, the Bay has black, egg-shaped berries which were used medicinally. For the scientific name see Ch.8 .

BERGAMOT (*Monarda didyma*). This herb would be best known by its other names 'Bee Balm' or 'Oswego Tea' since Bergamot is a thoroughly confusing one. It is the name of a pear from Turkish *beg armydu*, 'prince's pear', but at the town of Bergamo in Lombardy lemon oil was extracted from citrus fruit and sold as Essence of Bergamot and so Lemon Mint and other lemon-scented plants have also been known as Bergamot. *Monarda* is from Nicholas Monardes (1493-1588) a Spanish physician-botanist and *didyma* means 'in pairs', referring to the stamens, though it would apply equally well to the leaves.

BORAGE (*Borago officinalis*). Naturalised in Britain for many centuries, Borage has acquired native names like Bee Bread, Cool Tankard, and Ox-tongue. It has been prized for its supposed cooling and cordial effects and for giving courage. However, the Latin name *Borago* comes from two Arabic words meaning 'father of sweat' so it was evidently regarded as a febrifuge.

CARAWAY (*Carum carvi*). Caraway roots, according to Culpeper, are better food than Parsnips, but it is the flavour of the seeds added to a cake or a goulash that most of us know. Pliny, probably wrongly, associates the plant with the town of Caria in Asia Minor where he says it was grown, but it is indigenous to much of Europe and Asia. 'Caraway' appears to come from Arabic *karawiya* which in turn is probably from the Greek *karon* of Dioscorides and gives rise to both halves of the scientific name as well.

CHERVIL (*Anthriscus cerefolium*). The flavour is a delicate cross between Parsley and Anise and the name may mean 'leaf-of-greeting' or possibly 'waxy-leaf'. The problem is the 'cere-' bit of *cerefolium* which had become *cerefille* in Old English: is it from Greek *khairein*, 'to greet', or Latin *cereus*, 'waxy'? The experts I have consulted are divided. *Anthriscus* is from the Greek name for an umbelliferous plant.

CHIVES (*Allium schoenophrasum*). Culpeper accurately calls them 'Rush-Leeks' for *schoenophrasum* means 'rush-like'; he believes they send up hurtful vapours to the brain and spoil the eyesight. *Allium* is the old Latin name for Garlic and 'Chives' comes via Old French from Latin *cepa*, 'onion'.

SWEET CICELY (*Myrrhis odorata*). Add the leaves to rhubarb or sour fruit and you will need less sugar in the cooking. Try eating the green seed-pods raw too. The Greek *seseli*, an umbelliferous plant used in medicine, was confused with the girl's name, probably by old Gerard who seems to have been the first to spell it that way. *Myrrh* is a Semitic name for the precious gum-resin of the Bible, but was being used by the Greek Dioscorides for an aromatic plant with which the herbalists equated this one.

CORIANDER (*Coriandrum sativum*). Coriander was esteemed from very ancient times and once believed by the Chinese to confer immortality on those who used it. Geoffrey Grigson says that the Greek name *koriannon* 'is perhaps from *koris*, "bed-bug", on account of the scent and appearance of Coriander seeds'. I am pleased to report that I have never come within smelling distance of a bed-bug.

DILL (*Anethum graveolens*). In Old English it was *dile* and there are similar names in other Germanic languages of unknown origin. *Anethum* is from the old Greek name for the plant. Culpeper says, 'It stayeth the hiccough'.

FENNEL (*Foeniculum vulgare*). A decorative plant and every part of it is edible. The English name comes from the Latin one which means 'little hay' from the finely cut leaves. The variety *dulce* is the Finnocchio or Florence Fennel which has become popular here recently.

GARLIC (*Allium sativum*). In Old English *gar* meant 'spear', so Garlic is the 'Spear-Leek' from its lance-like foliage. Culpeper grudgingly admits it has some uses but says it has an offensive smell and adds fuel to the fire of choleric men. Until recently most Englishmen did not like the smell or taste any better than he did.

HORSE RADISH (*Armorica rusticana*). *Armorica* was the Latin name for a related plant and *rusticana* is 'of the country'. The English name arose from a mistranslation of German *meerettich*, which is literally 'sea-radish': see Ch.1.

HYSSOP (*Hyssopus offiicinalis*). It was *ysope* in Old English and the name comes via Greek from Hebrew *ezob*, 'holy herb'. It was used for cleansing lepers: 'Purge me with hyssop, and I shall be clean', sings the Psalmist (51, v.7), but it is thought by some that the Hyssop of the Bible was actually Marjoram.

LOVAGE (*Levisticum officinale*). I like the young leaves in salad but you could try putting them in your shoes to relieve tired feet. 'Lovage' comes by way of Old French *livische* from the Latin name *Levisticum*, which may in turn be from Liguria where Pliny says the plant grows wild.

MARJORAM (*Origanum vulgare*). *O. vulgare* is our native plant; *O. majorana*, Sweet Marjoram, comes from North Africa and Asia Minor and there are other species. *Origanum*, the Greek name, is from *oros*, 'hill' and *ganos*, 'brightness'. 'Marjoram' seems to come by way of French and medieval Latin from the *maezurana* of Dioscorides, but whether it is connected with the Romans' name *amaracus* is uncertain.

MINT (*Mentha*). Not a large genus but a confusing one because of the numerous hybrids and varieties which have been prized and cultivated around the world. There are about a dozen native species in Britain including Spear-Mint, Peppermint and Penny Royal. '*Minthe*' is the Greek name and perhaps varieties of *M. spicata* (spiked) and *M. raripila* (having scattered hairs) are the most commonly grown for kitchen use.

MUSTARD (White, *Sinapis alba*; Black, *Brassica nigra*; Brown, *B. juncea*). White Mustard is the fastest growing, the mildest and the one eaten with Cress as a salad. Its seeds are used in pickles.

Black Mustard has the fiercest taste. The Romans gave Mustard its name from *mustus* or unfermented grape-juice with which the seeds were pounded and *ardens* (from which our word 'ardent' comes) meaning hot or passionate.

ORIGANO, see Marjoram.

PARSLEY (*Petroselinum crispum*). As well as the familiar *crispum*, 'curly', there are flat-leaved varieties like Hamburg Parsley (v. tuberosum) and Italian or French Parsley (v. Neopolitanum). One of the few pieces of advice my father ever gave me was that a sprig of Parsley chewed would disguise the smell of alcoholic drink on the breath. The name comes by way of French and Latin from Greek *petros*, 'rock', and *selinon*, 'celery'.

ROSEMARY (*Rosmarinus officinalis*). Rosemary has only become a common name for girls in the twentieth century and not many of them know that it has nothing to do with Mary or Rose. *Ros nocturnus* in Latin is 'dew' and *ros marinus* is 'dew of the sea', in other words, I suppose, 'sea-spray'. I have certainly seen it looking happy in seaside gardens but it is not a native in Britain.

RUE (*Ruta graveolens*). It does have some culinary uses but is more of a strewing and medicinal herb with many supposed magical properties. It was *rude* in Old English which died out and our 'Rue' comes from Greek *hrute* by way of Latin and French and gave the Elizabethan poets an irresistible opportunity for punning (as did the name Thyme).

SAFFRON (*Crocus sativus*). The bulb was imported from Asia Minor but the Saffron grown in England was said to be 'the most excellent of all other'. It should not be confused with our native Meadow Saffron or Naked Ladies (C. autumnale), the source of a powerful poison. Unscrupulous cooks and apothecaries have used Marigold petals as a cheap substitute for Saffron. 'Crocus' is from Greek *krokos* and 'Saffron' comes by way of Latin from the Arabic name.

SAVORY (*Satureia montana*). This is Winter Savory, though *montana*, of course, means 'of mountains': there is also *S. hortensis* (of gardens) and *S. repandra*, Creeping Savory. They have been used for many purposes including the flavouring of salami. The Latin name in its Old French form *savriee* no doubt suggested to English ears 'savoury', i.e. 'tasty', though it has been suggested that it is connected originally with satyrs because of its supposed aphrodisiac properties.

TARRAGON (*Artemisia dracunculus*). *Dracunculus* means 'little dragon' and names of that kind indicate that the plant was once believed effective against snake bites, though it is now a culinary herb indispensable in Mediterranean cooking. 'Tarragon' comes from Arabic by way of Spanish and ultimately has the same etymology as 'dragon', while *Artemis* (or Diana to the Romans) was the virgin goddess of hunting and the moon.

THYME (*Thymus vulgaris*). We have three native species and the commonest is now called *drucei* after the Oxford botanist (see Ch.18) but there are many more species and varieties in cultivation. The name comes from Greek *thumon* meaning 'that which is included in a burnt offering or sacrifice'. It was also anciently believed to give courage and vigour to fighting men and to overcome shyness in girls. The down-to-earth Culpeper says it purges the body of phlegm and is a remedy for chincough in children and worms in the belly.

Strawberry
'Erdberen' from Hieronymus Bock's herbal of 1546.

Chapter 16
A Case of Fruit and Nuts

The earliest men were hunter-gatherers: they would have had more sense than to go gathering nuts in May, or fruit either, apart from an early gooseberry. I am all for nonsense verse, but what the nursery rhyme refers to are *knots*, posies or nosegays of such flowers as could be gathered on May Day, tied with a knot of ribbons (and, perhaps, grown in a knot-garden). So long as there were plentiful wild fruits in season there was no point in man cultivating them. What happened, we may suppose, is that pips and stones were thrown out around his settlements with the rubbish and germinated where the ground was disturbed and fertile. In time they produced fruits superior to the wild ones conveniently on his doorstep. So the first orchards were probably created by accident. It should not be assumed, however, that because the Strawberry and Plum, for instance, grow wild in Britain, they are the ancestors of our cultivated varieties or that they were domesticated here. The cultivated Strawberries actually have American parents, as we shall see, and Plums were developed in the Middle East.

So long as there have been gardens, fruit and nuts have been grown in them. The Romans brought many of their Mediterranean fruits here with them. Monastic houses with their contacts all over Europe also often grew non-native exotics like Figs, Grapes, Peaches and Mulberries. After the end of the Middle Ages, rivalry between great secular landowners and their desire to furnish their tables elegantly encouraged gardeners to experiment in growing and improving all manner of things, native and imported, especially from the Americas. The loss of the American colonies and the consequent high price of sugar meant that jams and preserves went out of fashion and fruit growing suffered a temporary decline. The early 19th century saw a strong revival and the Horticultural Society, founded in 1804, with its shows and prizes, encouraged great nurserymen like the Laxtons of Bedford and the Rivers in Hertfordshire to produce a stream of new and improved cultivars, many still grown today.

In the present century this work has been extended by the

R.H.S. at Wisley (founded in 1904) and by the research stations at Long Ashton near Bristol (1903) and East Malling in Kent (1913). The National Fruit Trials, begun at Wisley in 1922 were moved to Brogdale Farm in Kent in 1952. All this activity has led to the development of hundreds and, in the case of Apples, thousands of named varieties. We shall therefore have to limit our attention, as in other chapters, to the common and scientific names of the genera and species rather than those of the multitude of cultivated varieties.

ALMOND (*Prunus amygdalis*). The Greek name *amygdala* became *amandula* in late Latin, *amande* in Old French and so to English. We tend to grow ornamental varieties since our climate does not reliably produce useful nuts, but in Mediterranean countries one has the lovely sight of pink blossom on bare twigs in February followed by an important crop, even the shells making a splendid fuel.

APPLE (*Malus sylvestris ssp. mitis, M. pumila, M. communis*, etc.) The wild woodland (*sylvestris*) Apple is a native in Britain and not uncommon, being rather shrubby and having round yellow-green fruit which are no bigger than a ping-pong ball. *Malus* is the Latin name for the Apple and has nothing to do with the adjective meaning 'bad'. Our cultivated Apples are supposed to descend, in the main, from a subspecies of *sylvestris* called *mitis* meaning 'soft' as opposed to 'sharp' or 'hard'. There are other synonyms such as *pumila* (low, dwarf) and *communis* (common or gregarious, and there are whole woods of wild Apple trees in some places like Georgia).

Apples have been cultivated for at least five or six thousand years perhaps for cider-making rather than dessert originally, though there is archaeological evidence that Neolithic people sliced and dried the fruit for winter use. Today there are some 2000 cultivated varieties in the National Collection at Brogdale Farm and their names, which are even more numerous, would need a whole book to themselves. Some like Russet and Sunset are descriptive; some like Newton, Sturmer and Worcester are from places, and still others are named after people like Charles Ross, James Grieve and the Rev. W. Wilkes. There is a place called Bramley in Derbyshire and several other counties but the famous cooking Apple was raised from a pip by a

Miss Brailsford and planted in the garden of Matthew Bramley, a Nottingham innkeeper and butcher about 1850, so it does not do to jump to conclusions. Richard Cox was a brewer of Bermondsey who retired to Colnbrook where his Orange Pippin seedling was grown. The name 'Apple' has a long history (see p.20): there are cognate names in Germanic and Celtic languages, e.g. *Apfel* in German and *afal* in Welsh.

APRICOT (*Prunus armeniaca*). The Apricot's history is exceptionally complicated. In brief, it only grows wild in China where it must have been cultivated. It reached W. Asia and the Romans called it 'the Armenian tree' because that is where they had it from. They also called it *praecoqua* because of its *precocious* early flowering. The Arabs got hold of this name and it arrived in Spain with the Moors to become *albaiquoque*. French *abricot*, German *Aprikosa* and the Apricocks of Shakespeare's England all come from the Spanish or Portuguese names.

BLACKBERRY (*Rubus fruticosus*). The old name was Bramble, *bremel* in Old English, and from the same root as Broom. Many country names for the fruit like Brimmels or Brummel-kites are dialect variants of Bramble, though Bleggs or Blaggs in Yorkshire are presumably 'Blacks'. There are almost 400 hundred species of Rubus native in this country and the readiness of the Blackberries to hybridise with each other, with the Raspberry and the Loganberry make it a very complicated genus and gives us fancy fruits like the Boysenberry, Lowberry and Veitchberry. *Fruticosus* is nothing to do with fruit; it means 'shrubby'.

BLUEBERRY (*Vaccinium corymbosum, V. australe*). We have several native Vacciniums but none of them makes a satisfactory garden plant: there are Cowberries (*V. vitis-idaea*), Cranberries (*V. oxycoccus*) and particularly Bilberries or Whortleberries (*V. myrtillus*), known as Blaeberries in Scotland and as Worts or Hurts in the West Country. The name 'Bilberry' is said to be Scandinavian, but the origin of 'Whortleberry' is obscure. Tall American relatives have been developed from the two species named above, which make rewarding garden bushes if you have an acid soil: they turn splendid shades of scarlet and gold in autumn, as well as providing fruit which are much bigger than our wild ones.

CAPE GOOSEBERRY (*Physalis peruviana*). The *Physalis* genus is related to the Tomato and the name comes from Greek *physa*, a bladder. There are several species such as the Ground Cherry and Tomatillo which are grown for their fruit, but the Cape Gooseberry is the best known and most decorative with its lantern-like husk. As the specific name indicates, it came from S. America but is naturalised in S. Africa and widely cultivated there. The versatile berries can be eaten in either a vegetable or a fruit salad.

CHERRY (*Prunus avium, P. cerasus*). 'Cherry' comes by way of Old French *cherise* from the Latin *cerasus*, which is from an Asian name for the sour Cherry: it is also used as the generic name when the Cherries are given a genus to themselves. If you find it growing in a wood or hedgerow here, it is an escape from cultivation. The Gean or Wild Cherry, however, is a sweet Cherry and probably a true native. Its specific name *avium* means 'of birds' which is rather confusing since the Bird Cherry, another native, is *P. padus* (Theophrastus' name for it) with small black fruits which are bitter and useless. Tacitus reported in AD 79 that Britain produced good Cherry crops and 1900 years of cultivation have produced the 250 named varieties known today.

CHESTNUT (*Castanea sativa*). Some writers say that the Romans called the Chestnut after the town of Castania in Thessaly, but it may well have been the other way round, since Chestnut is *kastana* in Greek and there is a similar name in Armenian. Anyway the tree is native to much of S. Europe, N. Africa and S.W. Asia and was probably introduced to England by the Romans. It needs a good summer to crop well here and so imported 'Spanish Chestnuts' are mostly what we buy in shops. The Romans' name had become *cisten* or *cist-beam* in Old English which was later influenced by Old French *chastaigne*. If you know *Tristram Shandy*, you will make sure your dress is properly adjusted before eating roast Chestnuts.

COB-NUT (*Corylus avellana var. grandis*). '*Corylus*' is the Greek name for a large genus of which the English Hazel is one, and the Cob is a large cultivated Hazel. It is the only nut that has been grown on a commercial scale in England but it was never a very reliable crop. Old orchards of Kent Cobs, often interplaned with Damsons, were grubbed up during and after the Second World War. The name was originally 'cobil-nut' and a cob or cobble is a rounded stone, which the nut was supposed to resemble. Children used to play 'cobs' or 'cobnuts' before 'conkers' came along. Our other name, Filbert, we got from the Normans: St Philibert founded monasteries in Normandy in the 7th century and the nuts ripen around his feast day, August 20th.

CRAB APPLE (*Malus sylvestris*). They are 'Scrabs' in Northern dialects and the name probably came from Scandinavia with the Norsemen. There are 35 species, many grown for their ornamental flowers and fruit, but *sylvestris* is our native woodland Apple and the most important ancestor of our edible ones. Roasted Crabs were added to a bowl of mulled ale in Shakespeare's day (see Puck's speech in *A Midsummer Night's Dream* II, i.).

CURRANTS (*Ribes nigrum, R. rubrum, R. album*). A Tudor writer, Henry Lyte, calls the Red Currant a 'Bastard Corinth' reminding us that a Currant was originally a small dried grape or raisin from Corinth in Greece. '*Ribes*' is said to come from an Arabic or Persian name meaning 'acid tasting', but Currants are '*ribs*' in Danish and there are similar names in other Scandinavian languages. There are several species of Red Currant, two or three of which are native in Britain such as *sylvestre* and *petraeum* (growing in rocky places; you can find it in wooded valleys in the Peak District). Despite this, they do not seem to have been cultivated here until fairly recently: perhaps it was too difficult to keep the birds from making off with the crop. Botanically the White or Dessert Currant is not a separate species, only a variety of Red Currant. The Black Currant, *nigrum*, however, is a native of parts of Northern Europe and Asia and was unknown to the ancient civilisations. It was probably first cultivated in Russia and introduced to the rest of Europe by way of Germany but not widely grown until the 18th century.

DAMSON (*Prunus insititia*). The Damson was first called a Damasee or Damascene and was taken to be the *pruna damascena* of the Romans which, according to Pliny, came from Damascus. Bacon is calling them 'Damosins' in 1626. They are sometimes treated as a mere subspecies of the Plum, *P. domestica*: *insititia* means 'grafted' but none of my authorities reveal what was grafted on to what: it is the Cherry Plum, *P. cerasifera* (Cherry-like) which is most commonly used as a vigorous rootstock for cultivated Plums, but Damsons seed themselves readily and seem to do well on their own roots.

FIG (*Ficus caria*). The Fig genus includes the Banyan Tree and the Rubber Plant, *Ficus elastica,* (though commercially rubber comes from a different genus). The specific name *carica* is from the Province of Caria in Asia Minor and it is likely that Figs were first cultivated in that area four or five thousand years ago though where they originally grew wild is very uncertain: there is a theory that they only survived the last Ice Age in the Canary Islands. The Fig was *sukon* in Greece where the Athenians were notoriously partial to the fruit and an etymological curiosity is the way our word 'sycophant' meant 'fig-accuser'. In Rome Pliny knew 29 cultivated varieties and it was the Romans who brought Figs to England where they will only ripen out doors in the sunniest position and with root restriction.

FILBERT, see Cob-Nut.

GOOSEBERRY (*Ribes grossularia/Grossularia uva-cripsa*). The Gooseberry is a native plant all over Europe and was probably taken from the wild into gardens and improved. Gooseberry Clubs dedicated to producing giant fruit were once numerous and a few still exist. The Greeks and Romans do not seem to have been interested in a berry that was so obviously inferior to the Grape. The English name is probably a garbled version of the French *groseille* influenced by the fact that Gooseberry sauce goes well with roast goose. The scientists' *grossularia* seems also to be from the French which may ultimately mean 'crisp', while *uva-crispa*, its alternative Latin name does not mean 'crisp', but something like 'curly-bunched'.

GRAPE (*Vitis vinifera*). Wheat and vine, bread and wine are equally ancient products of man's ingenuity in exploiting nature. 'Wine' and 'vine' both derive from Latin *vinum* and before that there are related names in Greek, Hebrew and Armenian. 'Grape' is altogether different and has a more curious history, resembling that of 'spud'. It comes from French *grappe* meaning a bunch of grapes but, originally, 'that which was cut from the vine with a *grape*', a hooked knife: 'grapple' comes from the same source. The Domesday Book records 38 vineyards in William's newly conquered kingdom and if global warming continues the recent revival of interest in growing grapes in England seems likely to continue.

GREENGAGE (*Prunus domestica ssp. italica*). Greengages or *verdocchia* were introduced to France from Italy in the 16th century and called *prune de la reine Claude* by François 1st after his queen. A specimen was sent to Sir William Gage of Hengrave Hall near Bury St Edmunds in the reign of George 1st and, its label having been lost, so the story goes, it was named a 'Green Gage' by his head gardener.

HAZEL, see Cob-Nut.

LOGANBERRY (*Rubus loganbaccus*). The Loganberry was found growing wild in California in 1881 by Judge J.H. Logan. It was originally taken to be a natural hybrid between Raspberry and Dewberry but is now considered a red-fruiting variety of *R. vitifolius*, the Western Dewberry.

MEDLAR (*Mespilus germanica*). Medlars were well-known to Chaucer and Shakespeare. The polite name comes from the descriptive Greek *mespilon* (*mesos* = half & *pilos* = a ball). They are native to S.E. Europe and S.W. Asia and were probably brought to England by the Romans: however, there is reason to think that they were known to the Angles and Saxons too. For centuries the equally descriptive vernacular name was 'Openarse' and the same name is found in Frisian and German (*apenärseken*). 'Shit-snobs' was the name I knew them by as a boy in Kent. As a dessert, they are an acquired taste, rather like a woolly Pear to my mind. They have also been used to cure diarrhoea and, according to Culpeper, 'the longings of pregnant women for unusual food'.

MELON (*Cucumis melo*). The name comes by way of French and Latin from Greek *melopepon*, 'apple-gourd'. With the exception of the Water Melon (*Citrullus vulgaris*) which comes from Africa, all the cultivated sorts are varieties of this species native to S. Asia.

MULBERRY (*Morus nigra*). The Chinese White Mulberry

has tasteless pink fruit but silk worms relish its leaves. The Black Mulberry, however, is delicious but it is a frustrating fruit: it is very tart if not quite ripe and squashes when you try to pick it: as soon as it is ripe, it falls off and squashes again. The only solution is to lie hopefully under the tree - which can live to a great age - with mouth open. 'Mulberry' was earlier 'Morberry' and comes from the Latin and Greek names for the tree.

NECTARINE (*Prunus persica nectarina*). Nectar was the drink of the gods and etymologically it may mean 'triumph over death'. The smooth-skinned Peach was being called a Nectarine in 17th century England. Andrew Marvell in *The Garden* writes,

'The nectarine and curious peach
Into my hands themselves do reach.'

Who first gave it that name seems to be unrecorded but it is called a 'nectar-peach' in Dutch and German.

PEACH (*Prunus persica*). The ancestors of our Peach grow wild in parts of China and Korea, but the Romans had them from Iran where they were cultivated and so gave them the name *persicum malum*, 'Persian Apple'. It was *persoc* in Old English, but 'Peach' is from the Old French *pesche*. Several hundred varieties have been raised by nurserymen like the Rivers family of Sawbridgeworth in Hertfordshire. 'The fruit provokes lust', according to Culpeper, and the kernel bruised and boiled in vinegar 'marvellously makes the hair to grow again upon bald places': the downy skin of the Peach no doubt gave him this idea.

PEAR (*Pyrus communis*). The wild Pear, though less common than the wild Apple, is also a native here as it is in much of Europe and Asia. However, cultivated varieties probably originate in Asia Minor and were introduced by the Romans. A mural painting in the ruins of Pompeii shows them growing among other fruit in an orchard. Latin *pirum* became *pere* in Old English. Today many of our best varieties come from France or Belgium as names like Doyenne du Comice and Durondeau indicate.

PLUM (*Prunus domestica*). Ornamental Prunuses nearly all come from China or Japan but culinary Plums are more nearly related to our native Sloe and Bullace. The Romans probably had the Plum from Greece where it was *proumnon* and it may have been first domesticated there or somewhat further east in Syria or Iran. They introduced it to their colonies. In Old English the name had become *plume* which was also used for dried Grapes or raisins: 'Plum' pudding, I am assured, was never made of Plums or Prunes. Henry VIII had a Royal Fruiterer, one Richard Harris, who introduced new varieties from the continent: when Henry's warship, the Mary Rose, sank off Portsmouth Harbour in July 1545, it had five varieties of Plum on board which have been identified from their stones.

QUINCE (*Cydonia oblonga*). 'Quince' is really a plural of the earlier *coyn* or *quoyn* which comes via French from the Latin *cotoneum* or *cydoneum malum*, the 'Apple of Cydonia' in Crete (now called Canea). A ripe Quince is golden and highly aromatic and it is likely that the golden Apples that feature in Greek mythology were in fact Quinces. The fruit was dedicated to Aphrodite and featured in nuptial ceremonies. Marmalade is properly a Quince preserve, from *marmelo* in Portuguese.

RASPBERRY (*Rubus idaeus*). In England they were formerly calls Rasps or Raspis, a name of uncertain origin though a clue may lie in *vinum raspatum*, a sweet red wine imported from France and also known as 'raspis'. Raspberries grow abundantly wild everywhere in the cooler parts of Europe and Asia and were therefore not much esteemed. They are never mentioned by classical writers and have probably only been cultivated comparatively recently. The specific name *idaeus* means 'from Mount Ida': there is one in Crete and another in Turkey associated with the beautiful youth Ganymede who was carried off by Zeus to be cup-bearer to the gods - another link with wine.

RHUBARB (*Rheum rhaponticum*). The name has a somewhat complicated history and we do not eat the *fruit* of course: indeed, it was the root of the species *R. officinalis* which was made into a drug called *Rheum*. The Romans had it from China by way of Persia and we were still importing large quantities of it from China in the last century. Our culinary Rhubarb with the luscious pink stems is a native of Siberia but it was also grown originally as a medicinal plant. The Greeks called it *rha barbaron*, the *rha* of the foreigners or barbarians. Now *Rha* was the old name of the river Volga by which route it travelled south and the '*ponticum*' part of the specific name, meaning 'from the Black Sea area', points in the same general direction.

STRAWBERRY (*Fragaria vesca*). The ancestor of all our modern large-fruited Strawberries was produced in the 1820s when Michael Keen crossed a N. American Strawberry, *F. virginiana*, with a S. American one, *F. chiloensis*, to produce Keen's Seedling. Before that, our native *F. vesca* was cultivated in gardens, improving somewhat in size and juiciness, and is the ancestor of our Alpine Strawberries or *fraise des bois*. Another European Strawberry, *F. elatior* (taller), produced the once popular Hautbois varieties. The effusive Mrs Elton in Jane Austen's *Emma* (1816) thought them 'infinitely superior - no comparison - the others hardly eatable'. The Latin name is derived from *fragrans*, the fragrant fruit, and *vesca* means 'edible' as opposed to our other native, the Barren Strawberry, *F. sterilis*. The name 'Strawberry' goes back to Old English but may not be as obvious in meaning as it looks. Strawberries have their small seeds sticking out through the wall of the fruit and it has been suggested that these were thought to resemble chaff or fragments of straw.

WALNUT (*Juglans regia*). The Anglo Saxons recognised that the Walnut was not a native tree and called it *walh hnutu*, literally 'Welsh', i.e. 'foreign nut' (see p.27). It may be native to the Himalayas but was an early introduction to the Middle East and Mediterranean region. The Latin name is altogether grander: *Juglans* is composed of *Jovis* and *glans*, 'Jove's acorn', and *regia* is 'royal' or 'splendid'. It was a source of valuable dyes, medicines and timber as well as nuts, which were pressed for their oil

Chapter 17
A Bushful of Birds

As with plants, so with birds; that great Swede Linnaeus was the architect of our system of scientific naming. Each bird has two or often three Latin names for its Genus, Species and Sub-species. Thus the Blackcap is *Sylvia atricapilla* meaning, literally, the black-haired member of the Sylvia or woodland genus of warblers. The name 'warbler', incidentally, was coined by Gilbert White's friend and correspondent, Thomas Pennant in 1773. Our Goldfinch is *Carduelis carduelis britannica*: it has a genus and species all to itself from Latin *cardus*, a thistle, and its Old English name was 'thistle-tweaker', but the British Goldfinch is slightly different from the continental race, hence the sub-species name *britannica*.

English names, like Latin ones, pick up some particularity of the bird's appearance or behaviour in the same way as the nicknames we give each other. Blackcap and Goldfinch are appearance names like the surnames Blakelock or Golding which were originally nicknames; Wagtail and Treecreeper obviously refer to behaviour, Chiff-Chaff and Curlew to voice, Woodpigeon and Fieldfare to habitat and Mistle Thrush and Sparrowhawk to real or supposed food preferences. The Mistle Thrush certainly does like mistletoe berries.

The trouble with English names, picturesque though they often are, is that they vary from place to place and that can lead to ambiguity. Blackcap, for instance, is a name that has been given to the Bullfinch, Reed Bunting, Great, Coal, Willow and Marsh Tits and probably other birds, as well as to *Sylvia atricapilla*. And it, in turn, is known by a variety of English names such as 'Mock-Nightingale' which Gilbert White had heard it called in Norfolk. Its sweet inward song is as notable as its black cap (which the female does not have anyway) and it is also called Jack Straw and Hay Chat presumably from its chosen nesting materials.

Blackcaps are usually thought of as summer visitors but it is overwintering ones that sometimes come to my garden to feed on the pollen and nectar of winter-flowering shrubs like Mahonia. They hover like humming-birds within a foot of my window to reach the golden goodies. Goldfinches have also come to my garden because I have planted teasels to attract them. They have at least a dozen other names including Redcap and Seven-Coloured-Linnet, but I think of them as Teasel Finches.

All manner of bird rarities turn up from time to time in unlikely places but the Blackcap and Goldfinch would not ordinarily be thought of as garden birds. The twenty I have chosen to include in this chapter are likely to be more familiar to gardeners in town as well as countryside.

BLACKBIRD (*Turdus merula*). '*Turdus*' was the Romans' name for a thrush and the specific name *merula* seems to imply that the Blackbird is solitary, keeps itself to itself, as indeed it does outside the breeding season. The common name in Scotland and Ireland is Merle and comes from the Latin. The English name 'Blackbird' is obvious enough though the female is not black, but dusky brown with a lighter flecked throat and also lacks the cock's orange bill.

Shakespeare has Bottom sing of 'The ouzel cock so black of hue' and Ouzel or Uzzel survive as dialect names, along with Colley which means 'little coal-black one'. The saying 'to whistle like a Blackbird' means to do something effortlessly and one can forgive him much for his contribution to the dawn chorus but, in truth, he is not a blameless character. He is quarrelsome, noisy, chucks the gardener's mulch and compost around and his young are notably greedy in the soft fruit season, spoiling more berries than they eat.

BULLFINCH (*Pyrrula pyrrula*). The scientific name *pyrrula* comes from the Greek for 'flame-coloured' and the English name, as with bulldogs, bullfrogs, and bullrushes, suggests size and stockiness. Their stout bills tell you that they are seed-eaters but, unfortunately, when seeds are scarce in the early spring they will strip the buds from fruit-trees, hence names like Plumbird and Budfinch. Waste ground with docks and nettles or an ash tree with plenty of keys will keep them safely busy, however, and it pays the gardener not to be too tidy: they will not touch fruit buds if there is a neglected herbaceous border or dead biennials like Honesty and Foxglove about. Dead antirrhinums are another favourite and I have seen them hovering with surprising agility to reach the shrivelled berries on a honeysuckle. Also, in preference to any fruit-tree, they will go for the flower buds of the ornamental plum *Prunus pissardii* which is one good reason for giving it garden room; another is that it makes a superb climbing frame for the Clematis Nellie Moser. That good old Victorian naturalist, the Reverend C.A. Johns found that scattering dried peas in a lane adjacent to his garden was an effective device. One way or another, it should be possible to enjoy blossom, fruit and handsome Bullfinches in your garden.

CHAFFINCH (*Fringilla coelebs*). The name *Fringilla* is thought to be related to Latin *frigus*, cold, because Chaffinches were supposed to thrive in hard weather. Certainly, as with the Bullfinch, the male of the species never *looks* cold even if his head is blue and Bluecap is one of his local names. Grain, whether among chaff or stubble, is a favourite food but many different seeds and insects are taken and he is not wholly innocent of choosing buds or seedlings for dessert.

The curious specific name *coelebs* means unmarried because Linnaeus mistakenly thought that only the females migrated from his native Sweden. Other names are Whitefinch, Pied Finch, Fleckie and Sheelie (in Scotland): Spink, Chink, Pink and Prink all echo his usual call which sounds more like 'chwink' to me as I write.

The Chaffinch is such a familiar bird around our homes that it seems to have attracted very little in the way of legend or folklore. When Browning had nostalgic home thoughts from abroad, it was the Chaffinch singing on the orchard bough in April that he longed to hear. That vigorous, descending, musical rattle of a song has been rather successfully rendered as 'chip chip chip tell tell tell cherry-erry-erry tissi cheweeo'.

COLLARED DOVE (*Streptopelia decaocto*). Older bird books do not mention the Collared Dove which is now so familiar in our gardens. Half a century ago they were more or less confined to the Balkans and then, for some reason which no one fully understands, they started spreading west and north across Europe, first breeding in Britain in 1955. So far as I can see they do the gardener no harm unless he is driven berserk by their monotonous trisyllabic 'ku-KOO-kuk' which can sound deceptively like the cuckoo from the distance. Their courting display is more entertaining: they soar upwards, often clapping their wings above their backs, and then glide downwards with wings and tail spread uttering an amorous squawk.

The scientific name takes some sorting out. Greek *streptos* can mean 'twisted' as in *streptocarpus* which has fascinatingly corkscrew-like seed pods and *streptococcus*, a nasty twisted germ-baddie which can cause anything from a mild sore throat to scarlet fever. But *streptos* also means 'collar' and our otherwise rather plain little dove or *pelia* has a neat black collar round the back and sides of the neck. The specific name *decaocto* looks even more puzzling. The legend goes that a maid-servant was employed by her mistress for a miserly eighteen coins a year and she called on the gods to let the world know how ill-used she was. Zeus took pity on her and created the Collared Dove to shout 'decaocto' which anyone with a memory of school Latin will know means 'ten-eight'. A likely story!

GREENFINCH (*Carduelis chloris*). I have the impression that Greenfinches are bolder and more numerous than they used to be and the reason may be their partiality for peanuts. At one time they were more heard than seen. Their most characteristic call is a long drawn-out 'tsweee', like an unoiled hinge, which they utter monotonously from a high perch throughout the summer months and this gives them the name 'Peasweep'. Green Linnet is another of their names and, of course, their specific name *chloris* is from Greek for green (as in chlorophyl and chlorosis, the 'green sickness' suffered by adolescent girls and plants lacking iron). Though olive green is their predominant colour, they flash a pure gold rump as they fly away from you.

Recent research has shown that Greenfinches range over an area seven or eight miles in diameter and that up to a thousand individuals may visit one garden though only a few are present at any one time. Don't assume that the ones that are on your bird-table today will be there tomorrow. Take a close look at them.

HEDGE SPARROW (*Prunella modularis*). Dunnock, meaning 'dark-cock', is a better name than Hedge Sparrow, I think, because,.with its diet of tiny insects and seeds, it is clearly no relation of the House Sparrow. *Prunella*, a name the Dunnock shares with Selfheal, the little purple flower of our lawns, also means 'dark-coloured', and *modularis* means 'singing' or 'warbling'. Like the Wren, another unobtrusive little brown bird, the Dunnock has a rapid song which is surprisingly strong and penetrating.

Chaucer's name was Heysugge meaning 'hedge-flutterer' and referred, I suppose, to the Dunnock's wing-quivering courtship which is conducted, like all their affairs, on the ground. Other names are Shufflewing (for the same reason), Pinnock (which gives us a surname) and Hedge-Warbler. They have also been called Blind- or Foolish-Sparrows because they were thought to be imposed upon by the Cuckoo as foster-parents more frequently than other birds. Their own sky-blue eggs in their neat mossy nest are a striking contrast to the appearance of these dowdy but inoffensive residents of our gardens.

HOUSE SPARROW (*Passer domesticus*). 'Sparrow' comes from Old English *spearwa* and means 'flutterer'. One can sometimes hear them called Sparr in Kent and Sussex, Spug or Spuggy in the north and Spur in Scotland which comes from the Viking name *spörr*, while the playful Spadger does not seem to be regional. House Sparrows used to be called Thatch Sparrows for obvious reasons and it is hard to imagine where they nested before man provided the eaves of buildings for their convenience. Their stout bills tell you that they are seed-eaters like the Finches and it is only their liking for ripe grain that takes them far from our houses. They will congregate in the hedges around cornfields in large noisy parties and join flocks of finches in gleaning among the stubble after harvest.

Although he is alleged to have killed Cock Robin with his bow and arrow, the Sparrow does not seem to have had a generally bad character, except for being cheerfully noisy and untidy and occasionally squatting in the nests of Swallows and House Martins. True, Chaucer called him 'Venus' son' because he was regarded as of all birds the most lecherous and Sparrows' eggs were once regarded as aphrodisiac, something I have no wish to try. They will peck at unprotected seedlings in dry weather for the moisture and make dust-baths in freshly raked soil. I have also seen them tweak onion sets out of the ground when they are collecting nest-building materials, but the gardener has no option but to tolerate such minor misdemeanours.

JACKDAW (*Corvus monedula*). Jackdaws nest in holes in trees and on cliffs but old buildings and chimney stacks suit them well. They like company. If one comes down your chimney it is supposed to presage a death in the family; elsewhere it is probably thought a sign of good luck, such is the way with folklore. They are astute, watchful birds: as soon as I open my back door to throw some food scraps on the lawn they take off from the church tower and come gliding in. They stalk about glossily on the grass keeping a watchful forget-me-not-blue eye on the house. It was recently found that the chamber above the bell-loft was full right up with sticks from their nesting. It took two lorry-loads to remove them all and the Jackdaws have been expelled.

Both the Latin name for the crow family *Corvus* and 'Daw' come from their cawing and the Spanish, Italian and French all have similarly echoic names. Jack, as well as being a pet name for John, has a confusing number of different meanings: a jackboot is long, a jack-knife folds, a jackass is male (and stupid) and a jackpot is a large prize. However, the Jacksnipe is the smallest of the snipes and the Jackdaw is the smallest of the crows. Jack can also be a knave or rascal as in a pack of cards and it is not only at Rheims that Jackdaws have a reputation for making off with bright, shiny objects like metal labels, rings and coins. That is why the specific name is *monedula* from Latin *moneta*, money. A further possibility occurs to me: listen to Jackdaws and you will frequently hear them exclaim 'tchak'. Other country names are Jackie, Grey-neck and Cadder from the old name Cawdaw.

MAGPIE (*Pica pica*). Magpies used to be persecuted woodland birds like Jays but they have now moved into our towns and cities and become the gangsters of the suburban garden, the bandits of the bird-table. I have seen them nesting on a roundabout at a busy junction and feeding their young in a city car park with no tree or blade of grass in sight. They will even defy the traffic with their sidling hops to scavenge the bird, insect and animal casualties on our roads. They may be 'the handsomest and most engaging members of the crow family', as one of my books puts it, .but they have always had a reputation for thieving and hoarding and have even been called 'the Devil's bird'. This is because they were variously alleged to be a cross between a Raven and a Dove, to have refused to go into the Ark or to put on full mourning at the time of the crucifixion. In Shropshire once it was the custom to take off one's hat, spit in the direction of a Magpie and say 'Devil, devil, I defy thee'.

Just as the pie you eat originally contained a mixture of things so 'pied' meant a mixture of colours. Shakespeare has 'Daisies pied ...' in the lovely song that ends *Love's Labours Lost*. Only later did 'pied' come to signify black-and-white. Shakespeare also has 'chattering pies' in *Henry VI* and the oldest English name for the Magpie is Hagister which means 'chatterer'. The Latin name 'Pica' echoes one of their exclamations. 'Chatternag' and 'Long-tailed-Nan' are other local names but Mag and Maggie, which are short for Margaret, is what the Pie has become for most people.

NUTHATCH (*Sitta europea*). I spun a coin to decide between including the Nuthatch or the Heron in this chapter. Herons seem to have survived loss of habitat in the countryside remarkably well and the reason may be the popularity of ornamental fish-ponds in suburban gardens which are robbed, very often in the early hours before anybody is about. However, the Nuthatch, meaning 'nut-hacker', won. Whether they are regular visitors to your bird-table will depend on the presence of mature trees in the vicinity. A town park, churchyard or cemetery often has one or more pairs of resident Nuthatches. *Sitta* comes from the Greek of Aristotle for a bird that lives by pecking the bark of trees and Woodhack is one of their names. What they are after, of course, is the grubs and insects that live in or under the bark but I have often seen them going to and fro from top to bottom of my garage roof as systematically as a knitter, probing under each tile. Acorns and hazel nuts are a bonus in autumn and they are superbly adapted to hack them open having first wedged them in a convenient crevice. When a pair come to my nut cage, they always take turns, one waiting in a nearby tree till the other has extracted a kernel and flown off with it. Jobbin or Nut-Jobber are other country names. They are unlike any other British bird in preferring to feed and explore among the trees head downwards and they are unlike the woodpeckers in not using their stumpy tails to sit on. They do, however, make woodpecker-like tappings and other sounds, notably a 'be quick, be quick, be quick' call and a rapid sequence of notes like a pebble bounding over ice. Referring to this call, I suppose, Gilbert White's Hampshire parishioners called them Jar-birds. They make their nests of papery scraps of birch bark in the hole of a tree and their other names of Mud-dauber or Mud-stopper refer to their habit of plastering round the edges of the hole until it is the right size to suit them and keep out larger intruders.

ROBIN (*Erithacus rubecula*). 'Robin' has only been officially accepted by ornithologists as an alternative name for the Redbreast since 1952. The Old English name was *Rudduc* meaning 'red-cock' and both halves of the Latin name also refer to the Robin's most conspicuous feature, its breast which is really a rich orange rather than the scarlet of the picture books. The male and female are impossible to tell apart in appearance. Early postmen wore red waistcoats and were inevitably nicknamed 'Robins' by the public.

The Robin is often called tame or confiding and Robin is an affectionate pet-name for Robert. When it was first tacked on to Redbreast in the late Middle Ages, the most familiar bearers of the name were Robin Hood and Robin Goodfellow who were bold and spirited, not a little contrary and even capricious but certainly not *tame*. The story of the Babes in the Wood who were covered with leaves by the Robins to keep them warm belongs to an altogether more tame not to say soppy period of history. In fact Robins are notably pugnacious in defending their territories and battles to the death have been recorded. We flatter ourselves, too, if we think they need human company for they are perfectly at home in the Greenwood and even in the remotest conifer plantations of the Forestry Commission. When they do take up residence in our gardens they are often too fearless for their own good: many a time they are narrowly missed by the spade when they dart between the digger's legs to seize some wriggling morsel. They do sometimes take the odd currant or raspberry as well, but I cannot imagine the gardener who does not admire their sprightliness and cheek. On the continent where little birds get eaten they are noticeably more cautious but in England they seem to lead charmed lives.

SONG THRUSH (Turdus philomelos). The Latin name of the thrush family has an unfortunate sound in English and the authorities seem divided on the specific name between *ericitorum*, 'of the heath' and *philomelos*, 'song-lover'. I know which I prefer. It was, after all, the Thrush's 'carolings of such ecstatic sound' on a winter's afternoon which inspired one of Hardy's best known poems, and Browning in Italy imagined the wise Thrush on his blossoming pear-tree back in England singing each song twice over. Indeed, the Thrush's very variable song does typically consist of repeated three- and four-note phrases which have suggested 'Shut the gate' and 'Did he do it?' and, no doubt, other catches. He is also, like the Starling, a good mimic of other birds.

Names by which the Song Thrush have been known include Throstle, Thirstle, Dirsh and Mavis, which was first used as a girl's name by Marie Corelli in one of her romantic novels. She also invented the name Thelma. In East Anglia Mavis is the name for the Song Thrush and Throstle for the Mistle Thrush but in parts of Scotland it is the other way round: that's the problem with non-scientific names.

As a nest-builder, the female Thrush - for she does all the work - is an expert plasterer. The very durable cup with which she lines her nest is not made of mud but usually of a mixture of rotten wood pulped with cow-dung and saliva.

Every gardener knows that thrushes are partial to ripe fruit and ornamental berries, but they also eat a prodigious quantity of snails which are smashed out of their shells on a favourite anvil stone. And who would be without that song?

STARLING (*Sternus vulgaris*). The gregarious Starling makes itself at home almost everywhere from the city centre to the farthest Hebrides. Just as a duckling is a little duck, so Starling was originally diminutive of Stare, the old name for the bird which was still preferred by eighteenth century naturalists like Thomas Pennant.

The German and Scandinavian names are similar and the Latin *Sternus* is not far removed which indicates that it is a very ancient name indeed, whose etymology is lost in the mists of prehistory. It is also the only bird in this chapter for which I know of no other English name except Starnel and this is surprising for it is glossily handsome, quite a character and happy enough to nest in any man-made hole or corner it can find. The nest is an untidy affair and the female untidily quite often drops an egg in mid air before she can get there. They are very good mimics and their 'song' is a cheerful medley of whistles, clicks, rattles and chuckles. In spring, particularly, their plumage has a showy pearl-sewn metallic iridescence of greens and purples. Their offspring are dusty brown but in late summer the adult plumage often appears in irregular patches giving them a comic piebald appearance.

A flock of Starlings can strike terror into the heart of a professional fruit-grower but the small parties that visit our gardens generally confine their attention to probing the lawns for worms and leather-jackets. They run or walk and are a considerable step up the evolutionary ladder from birds which merely hop. In winter they gather in huge numbers: one of the most awe-inspiring sights of the countryside is a flock of countless scores of black dots ravelling and unravelling against a frosty sunset, wheeling and turning with perfect synchronization until they suddenly drop to roost in some isolated clump of trees.

SWALLOW (*Hirundo rustica*). HOUSE MARTIN (*Delichon urbica*). The Swallow and Martin are everywhere welcomed as harbingers of spring. Though they fly, feed, build, breed and eventually depart in similar fashion, there is really no excuse for confusing them. The Swallow has much the longer tail as well as a red chin and the House Martin's white rump distinguishes it from both Swallow and Sand Martin. Their names, however, do suggest some confusion between the two species in the past or, at best, a recognition that they are closely related. *Hirundo* is from the Latin and *Delichon* an anagram of *Chelidon* from the Greek for Swallow; (Celandine or 'Swallow-wort' comes from the same source). The name Swallow is from an old Germanic word which means literally a forked stick, while Martin may be from *Martinette* which is the French name for the Swift, or perhaps because it was supposed to depart around Martinmass, November 11th, and may actually have done so under the old calendar. Shakespeare's King Duncan in *Macbeth* called it 'the temple-haunting martlet'. Before man provided temples or humbler structures for them to build on, and overhead wires for perching, they were birds of cliffs and caves. Although *urbica* means 'of the city', the Martin likes open spaces and water to skim over as much as the Swallow. It has been known as the House-, Window- or Eaves-Swallow and is the better architect of the two, sticking its mud cup on the sheer wall. The Swallow sometimes does the same but prefers to use a ledge or rafter in a porch, barn, stable or other outhouse shelter. It has been called the Chimney-Swallow though I doubt whether modern chimneys provide suitable building sites. Both are birds of our buildings rather than our gardens for they seldom alight on the ground except to collect mud for their nests. Both are also birds of good omen though there is in some places a superstition that if they fly between the legs of a cow it will become sterile. Gilbert White noticed that in autumn great parties of Swallows gathered to roost in reed beds and concluded that there might be some truth in the belief that some of them hibernate in the mud at the bottom of ponds.

GREAT TIT (*Parus major*). BLUE TIT (*Parus caeruleus*). COAL TIT (*Parus ater*). The three Tits most likely to be seen in our gardens are very similar in feeding and nesting habits and general behaviour though the Coal Tit is somewhat shyer and therefore less well known than the other two. Tit is short for Titmouse which is the old name for the whole family, but has nothing to do with mice (or with tits in the vulgar sense). Tit comes from Old Norse *tittr* meaning small and 'mouse' in this context is Old English *mase* meaning a small bird. Their Latin names are straightforward: *caeruleus* is sky-blue and *ater* black. In English both Great and Blue Tit are known as Tom-tit.

The Great Tit has the additional names of Ox-eye, Black-cap, Saw-finch or Saw-sharpener. The last two come from its repetitious, rasping 'tee-chew', tee-chew' call which has been thought to sound like a saw being sharpened with a file. It is a surprise to many people to learn that the Great Tit is not merely pugnacious but has carnivorous tastes and will kill and eat the brains of weakly small birds and take the fledglings from other birds' nests. The Reverend C.A. Johns tells of one which was mistakenly placed in a well-stocked aviary and killed all the other occupants in a single night. They also have the bad habit of pecking ripe apples and pears, thus allowing wasps to start their excavations, but when they strip buds from trees and shrubs in spring I believe they are searching out insect pests.

The Blue Tit can also be an irascible little bird which will raise its blue crest and scold anyone who interrupts its busy life and hiss and peck at the hand that comes near when it is sitting on the nest, hence the schoolboy name of Billy Biter. Other names are Blue Yaup, Blue-cap, Blue-bonnet, Pick-cheese and Hickmall.

Coal Tits, like Goldcrests, seldom go very far from conifers and their numbers fluctuate a lot depending on the severity or mildness of our winters. Coalmouse (or Colemouse) is the only other name I know and it is surprising if there is not a country name that refers to this rather sooty little bird's most distinguishing feature, the broad white stripe on the back of the head and nape.

WILLOW WARBLER (*Phylloscopus trochilus*). The Willow Warbler is much the commonest of the slim little greenish-yellowish migrants to reach us from Africa in early April. I have a large willow tree and they certainly spend much time exploring it for insects but they are not fussy and any garden with trees and shrubs is likely to be home to one or more pairs during the summer months.

Of the Latin names, *Phylloscopus* means 'leaf-explorer' and that is spot on because it is exactly what they do with great agility, occasionally chasing small moths on the wing as well: *trochilus* seems a less happy choice to me. It comes from a Greek verb to run and the Willow Warbler doesn't run in the way that a Wagtail or a Blackbird will. Perhaps Linnaeus had in mind their constant movement and restless energy among the branches.

Apart from Willow Wren and Haybird (from their supposed liking for the vicinity of hay fields) there don't seem to be any other English names for this little bird. This may be because of their inconspicuousness among the foliage and the notorious difficulty of distinguishing them from the Chiffchaff and several other warblers except by their song. This is unmistakeable and often repeated several times a minute. It is rather well described in one of my books as 'a liquid, musical cadence, beginning quietly and becoming clearer and more deliberate, *descending* to a distinctive flourish 'sooeet-sooeetoo''. There is also a plaintive call of 'hweet' usually uttered when they are well hidden and difficult to locate.

WOOD-PIGEON (*Columba palumbus*). When Britain was covered with oak forest and the Wood-Pigeon was perforce a pigeon of the woods, it filled its large crop with acorns, beech mast, haws and the young leaves of woodland trees. But when man cleared the forest it turned its attention to his grain and brassica greens and became a handsome villain.

Ring-Dove was the common name for the Wood-Pigeon for centuries and you find the same name in German and Dutch. Actually the Wood-Pigeon has white *patches* on the sides of the neck and Ring-Dove makes a better name for some of the domesticated pigeons which do have a dark neck-ring.

The Latin *Columba*, just meaning 'pigeon', gave us the Old English name Culver and a culverhouse, which survives as a surname, was the dove-cote. 'Pigeon' is said to come via French from Latin *pipio*, to chirp, which seems absurd on the face of it. All the other names for the Wood-Pigeon such as Clatter-Dove and Too-zoo refer to the noise it makes, particularly its five-note 'song' which has been thought to sound like 'Tak' two coos, Taffy'. In Scotland and the north Cushat, Coo-shot and Cow-shoot are common and probably mean 'coo-shouter'. Queest or Queech which one meets in the west from Cheshire and Shropshire down to Devon may be just variants of Cushat but I have seen it suggested that they come from Latin *questus*, a lament or complaining. However, it is the gardener who does the lamenting if he neglects to net his sprouts before hard weather and hunger make the Wood-Pigeons bold.

WREN (*Troglodytes troglodytes*). The Wren has collected about it much folk-lore and not a few different names including Our Lady's Hen. There are curses on anyone who kills a wren but paradoxically it was also hunted and killed on Boxing Day because it was supposed to have alerted the guards to St Stephen's escape and so brought about the death of the first Christian martyr. No doubt this was an appropriation by the Church of an earlier pagan custom in which, perhaps, the Wren with its earthy colours was seen as the spirit of winter, which had to be killed before spring could be born.

The name Cutty, Scutty, Stumpy and Bobby suggest the Wren's jaunty tail and habit of bobbing up and down before or after flinging out its rapid, prolonged, breathless rattle of a song. It is also known as Kitty-, Chitty-, Puggy-, Juggy-, Sally- or Titty-Wren and, I read, as Crackadee and Gulliver. Whether 'Crackadee' refers to the Wren's song or its 'tick-tick-tick' alarm call or its angry churr I can't guess, but the name Gulliver, coming from Old French for 'glutton; may have been suggested by the Wren's rotund appearance. As they nip about with ceaseless energy like hyperactive mice, they take large numbers of aphids and other small creatures which must be good news for the gardener.

The Latin name *Troglodytes* means 'cave-dweller' or 'hole-frequenter' and the Wren certainly likes to place its domed nest among rocks or in a stone wall. In a memorable passage in *The Prelude*, Wordsworth recalled hearing a Wren sing in the gloomy ivy-clad ruins of Furness Abbey. The cock bird often builds several nests, one of which the hen chooses to line with feathers and lay her eggs in. In hard weather they sometimes congregate to roost together: one winter morning at sunrise I saw eight emerge from the nest-box on my garage wall and much larger numbers have been recorded.

CHAPTER 18
A BUNCH OF BRITISH BOTANISTS

There is one simple fact about English surnames: they can all be fitted into one or other of four broad categories. Most numerous are the surnames from place-names or localities, then come the first-name surnames like Adams or Johnson, followed by occupational names and finally nickname surnames. Of course, there are always obscure names whose meanings can only guessed at and, also, the names of denizens: Patel and Singh are common names in Britain today but, like Jewish or Huguenot French names, they are not really part of our ethnic stock of surnames which go back to the Middle Ages.

We can illustrate two and, perhaps, three of these categories with the name of the Rev. Keble Martin (1877-1969). Martin was a very popular first-name after Saint Martin who, as a young Roman soldier, gave half of his cloak to a naked beggar and later became Bishop of Tours in Gaul. Keble, however, comes from the Old English word *cybbel* meaning a cudgel and might have been an occupational name for a maker of such weapons or, perhaps more likely, a nickname for a man whose head looked like a club or who was handy at using one, we can't be sure.

Another clergyman-botanist (and a superb poet), Andrew Young (1885-1971) provides another example of a nickname surname: to read his two books *A Retrospect of Flowers* and *A Prospect of Flowers* is almost as delightful as it would have been to share the plant-hunting excursions he describes. Philip Miller (1691-1771) of *Gardener's Kalendar* fame and Sir James Edward Smith (1759-1828), founder of the Linnean Society, provide two examples of common occupational names. For local surnames, we could take Sir Joseph Banks (1743-1820) who sailed with Captain Cook and was President of the Royal Society for forty-two years and George Bentham (1800-1884) who is indissolubly linked with Joseph Hooker (see below). Banks lived 'at the bank' and Bentham is a Gloucestershire place name meaning 'ham with bent-grass'. Sir A.G. Tansley (1871-1955) has the name of a Derbyshire village meaning lea or clearing in a branch- or side-valley. His monumental *The British Isles and their Vegetation* was published in 1939.

Now let us add a dozen more to our bunch of British botanists, in chronological order. Their names will be already familiar to anyone interested in plants, but the meanings of those names may be less so.

WILLIAM TURNER (? - 1568). Called the father of British botany, he published his *New Herbal* in 1551 which described plants from observation rather than classical sources. He was the son of a tanner of Morpeth, went to Cambridge to study medicine and became a passionate follower of the reformation doctrines of Ridley and Latimer. This necessitated periodical flights abroad to avoid sharing their fate at the stake with the incidental advantage that he met continental botanists.

But what did a turner turn? The answer is, usually a lathe, sometimes a spit, possibly even his coat or his tail. He may also have been one who jousted at a tourney, or a translator - very necessary in medieval times when English, Norman French and Latin were all in use for different purposes. There is even evidence that Turner has absorbed the nickname 'turn-hare' who was presumably so swift he could outrun a hare just as Turnbull was strong enough literally to take the bull by the horns.

JOHN GERARD (1545-1612) was born at Nantwich in Cheshire, became a successful barber-surgeon and cultivated over a thousand species in his garden at Holborn. His *Herball or Generall Historie of Plantes*, 1597, is not impressive as a work of original scholarship, indeed it was largely pirated, but his picturesque descriptions of plants and their supposed medicinal properties has made it the most famous English herbal. Thomas Johnson brought out a greatly improved second edition after Gerard's death.

Gerard is an old Germanic name introduced to England by the Normans meaning 'spear-bold': the first element appears in some plant-names like garlic, the 'spear-leek'.

JOHN PARKINSON (1567-1650). His fame rests on two books and an ingenious pun on his own name. The books are his *Theatrum Botanicum* which attempted a complete account of the plant world, and a horticultural treatise, *Paradisi in Sole Paradisus Terrestris*, meaning Park-in-sun's Earthly Paradise. Whether he believed that is what his name meant, I do not know, but Parkin or Perkin are contractions of the pet name Peterkin, so Parkinson means 'son of little Peter'.

NICHOLAS CULPEPER (1616-1654). He got a good knowledge of Latin and Greek at Cambridge, fought on the side of Parliament in the Civil War, receiving a wound in the chest leading to a consumption which shortened his busy life. He was not really a botanist, rather a rogue physician and self-proclaimed astrologer of Spitalfields. He became famous and wealthy through his cures, though he was reviled by the medical establishment. His famous Herbal has probably never been out of print and he left more than seventy other works, many of which were published posthumously by his widow.

We are told that Culpeper means 'cull (gather) pepper' and is the nickname for a spicer, but what pepper would a spicer have gathered in England I wonder? He describes water-pepper as growing three or four feet tall, so it can't be the little polygonum with the hot taste I know. His pepperwort or dittany is better known as Horse Radish: boiled in ale and drunk by an expectant mother, it will hasten her delivery. One of his works, incidentally, was a treatise on midwifery.

JOHN RAY (1627-1705) was the brilliant son of a blacksmith of Black Notley, Essex. He was sent to Cambridge and became a clergyman and a Fellow of Trinity. He was forced to resign his college appointments in 1662 by the Act of Uniformity which was designed by revengeful royalists to oust puritans from the Church. Freed from official duties, however, he was able to make botanical expeditions all over the British Isles and produce his great *Synopsis Stirpium Britannicarum* in 1670 - in effect the first British flora. Many rare plants can still be found growing in the places where he first identified them.

The name Ray has two possible explanations: it is either the Scottish and Northern form of roe (-deer) or from Old French *rei* meaning 'king', probably a nickname denoting pride or regal bearing or from having played the part of king or been elected 'king' of some feast or festivity.

JAMES SOWERBY (1757-1822) drew the beautiful illustrations for the *English Botany* which came out in thirty-six volumes between 1790 and 1814 and is always known as Sowerby's Botany. In fact the text was by Sir James Edward Smith who, as a young medical student, spent a thousand of his father's guineas acquiring Linnaeus's library and herbarium. The Swedes realised too late what a treasure they had lost. An engraving published in 1800 shows a Swedish warship chasing the English brig *Appearance* with twenty-six large packing-cases aboard across the North Sea, though apparently this incident never really occurred.

Sowerby is a place-name in several northern English counties and comes from Old Norse meaning a muddy or dirty place.

SIR WILLIAM HOOKER (1785-1865) and *SIR JOSEPH HOOKER* (1817-1911). The Hookers, father and son, were successively directors of Kew Gardens for almost half a century. Sir William was an administrative genius as well as a great botanist and established Kew's international reputation. His own collections formed the basis of the library and herbarium. Joseph carried on his father's great work but was also a tireless traveller and plant collector and also Darwin's closest friend and helper. According to Wilfred Blunt in his delightful history of Kew Gardens, Mea Allan's book *The Hookers of Kew* was bought by many subsequently disappointed Americans under a misapprehension about the meaning of the title. A hooker, however, was either a maker of hooks or an agricultural worker who used a hook or sickle - a reaper in fact.

CHARLES DARWIN (1809-1882) would have been famous as a botanist even if he had never written a word about the origin of species or the descent of man. He did important experimental work on the fertilization of orchids, on insectivorous plants and many other topics.

Darwin is an Old English name which means 'dear friend' just as Godwin and Baldwin mean 'good friend' and 'bold friend'.

GEORGE CLARIDGE DRUCE (1850-1932) was the illegitimate son of a farmer who was largely self-educated and qualified as a pharmacist. He saved for many years to buy a chemist's shop in Oxford where he prospered, becoming mayor in 1900. His *Flora of Oxfordshire* gained him an honorary degree from the University and access to its libraries and herbariums. From 1903 until his death he was secretary of the British Botanical Exchange. As a schoolboy during the 1940s I found a mistake in Hayward's *Botanist's Pocketbook* which Druce had revised several times, and wrote to him pointing it out. After many months my letter was returned, the envelope almost obliterated with redirections and postmarks. His London club had scrawled 'obit 1932' on it.

Druce (or Drewes) is a Norman name probably from Dreux in the Eure-et-Loire region of France.

C.T.&W. For nearly half a century now, these initials have been used to refer to the botanists' bible, Clapham, Tutin and Warburg's *Flora of the British Isles*, 1952. They were all Cambridge trained botanists: A.R. Clapham b. 1904 became professor at Sheffield from 1944-1969; T.G. Tutin, b. 1908, at Leicester; and E.F. Warburg, 1908-1966, who was known as 'Heff' (for Heffalump) because of his size, was curator of the Druce (q.v.) Herbarium at Oxford.

Clapham is a place in Bedfordshire which, like Clapton and Clapcot, contains an Old English word meaning rocks or hillocks. Tutin comes from a Scandinavian name Thurstan, 'Thor's stone', which became popular in Normandy as Toustain and hence Tustin, Tutin and other spellings in English. E.F. Warburg was actually of German-Jewish descent but the name can also be English: Warborough in Oxfordshire was Wareburg (lookout-hill) in the thirteenth century and there was also a St. Werburg, a royal abbess and daughter of the Mercian King Wulfhere, whose shrine was at Chester until it was destroyed in the time of Henry VIII.

John Parkinson 1567-1650
from the title page of his *Theatrum Botanicum*, 1640.

Chapter 19
The Name on the Gate

At the end of our lane when I was a child, the pub was called The Bird in Hand. The two old spinster sisters who lived opposite had a neatly trimmed holly in the front garden and called their cottage 'Two-in-a-Bush'. Not far away, the great Kent and England cricketer, Tich Freeman, lived in retirement and his bungalow was called 'Dunbolin'. We thought all this was marvellously witty.

I now live in a village in a cul-de-sac leading to the church which contains all of four houses. Recently a Council official wrote to say that we were to be numbered and asking where we wished the numbers to be 'affixed' to our properties. We refrained from telling him what he could do with his numbers, and one afternoon a lorry arrived with two men and a boy. The two men watched while the boy did the 'affixing' thus, at the turn of a screwdriver, making our house-names redundant. Before the numbers came, the names were useful and even necessary, and the purely descriptive names were the most useful, The Corner House, The Old Vicarage and so on. Anyway, the names have survived: we seem attached to them.

Maybe men have given names to their houses ever since they stopped living in caves. Frederick the Great's summer residence was called 'Sans Souci', Gibbon wrote *The Decline and Fall* at 'La Grotte' and, to go from the consequential to the comic, those cartoon characters of my childhood, Pip, Squeak and Wilfred, lived at 'Fair View'. But it is urban man and particularly suburban man who most seems to need to express himself in this way. We became a nation of city-dwellers in the second half of the nineteenth century when the population rose from thirteen to thirty million. The straighter the city street and the more uniform the houses, the more fanciful the names are likely to be: 'Costa Brava' - perhaps the location of a honeymoon or a successful holiday - has as neighbour 'Kostalot'; 'Skynt' is next door to 'Stony Broke'; 'El Nido' is by 'Kosinouk' and 'Casa Mia' and 'Cartref' (Welsh for 'home') beside 'Weeowna' and 'Ourome'.

Then there are his-and-her names like 'Ronkath', 'Phylbert', and 'Maveric' (presumably Mave and Eric rather than the eccentric Texan rancher Samuel Maverick who gave his name to the language). I have even heard of the unfortunate 'Renal' which must be for Rene and Albert unless a kidney specialist lived there. There are back-spellings like 'Tivuli', 'Emoclew', 'Llawnroc' and even 'Llamedos', and there are in-jokes like 'Hindquarters', the house of the Hinds, 'Coppernobs' for the Redheads and 'The Keep' for the Norman family's modest home. The couple at 'Two Hoots' seem to be telling the world of their nonchalance or waggishness. 'Cobwebs' looks rather twee until one is told it stands for 'Currently owned by Woolwich Equitable Building Society'. Beside all these, 'Tusikso' (260), 'Fortitoo' or '7777777' for '49' seem uninventive and the temptation to call 2B 'Ornot' should be resisted: it has been thought of too often.

All this humour and ingenuity are not, however, the commonest ingredients of house-naming, but rather nostalgia, nostalgia for a more rural past. The city dweller who can afford to do so attempts to corral a little bit of *rus in urbe* around him and chooses a name to match. Civic authorities, dreaming of garden cities, put flowerbeds on traffic islands, hanging-baskets on lamp-posts and express their frustrations in their choice of names for new roads. At one time they were often named after generals or battles, statesmen or local dignitaries, but this rarely happens today. It would be nice to think that some great gardener or plantsman has been honoured in this way, but I do not know of an example. Today even 'road' and 'street' tend to be avoided and 'Place', 'Terrace', 'Crescent' and 'Square' are also felt to be too urban and geometrical. Far more popular are 'Drive', 'Avenue', 'Way', 'Grove', 'Close' and 'Ride', frequently coupled with a name giving an ambience of spurious gentility or even grandeur such as 'Manor Way', 'Chatsworth Avenue' or 'Windsor Drive'. Archaic words like 'garth', 'royd', 'dene' and 'leaze' are sometimes resurrected in both street and house-names and given a countrified prefix: 'Oakhill' or 'Orchard' or 'Meadow' are very acceptable, but not 'Midden', 'Cabbage Field' or 'Silage Pit': a whiff of the real countryside is not welcome in sanitized subtopia.

House names follow the same trend. One can be fairly sure that 'Wychwood', 'The Briars', 'Merrydown' and 'Tanglewood' are in tamed and tidy suburbs, that 'Larkrise' knows only domestic sparrows and that 'Sunnymead' and 'Cloverleas' are in a concrete jungle many miles from the nearest meadow. Plant names are perennially popular with the Rose as the all-time favourite in umpteen compounds, and Jasmine, Hollyhock, Foxglove, Edelweis, Bluebell and Daffodil all high in the league table. I dare say that most plant-names from Aaron's Beard to Zygocactus have appeared on a garden gate somewhere at some time, though some like Bloody Burnett, Bladder Campion and Stinking Hellebore seem unlikely candidates, while Schizanthus and Streptocarpus - beautiful though they are - sound too much like nasty diseases. A name like 'Miraflores' saves the trouble of being specific.

Tree-names are, perhaps, even more popular than flowers and are sensibly descriptive provided the species named really is in evidence in the garden. 'The Elms' often survives long after the trees have succumbed to Dutch Elm disease and been replaced by Leyland's ubiquitous Cypress or what that great tree-man Alan Mitchell calls 'Pissard's awful purple Plum'. Suburban England looks all set to become a dense forest of 100-foot Cypresses early next century. Cedar and Acacia seem to be favourite aliens with house-names while most of our native species, Oak, Beech, Birch, Lime, Ash and Pine are frequently seen in compounds like 'Ashleigh' or 'Oakdene'.

Devotees of *The Diary of a Nobody* will remember that the Pooters (including son Lupin) lived at 'The Laurels' and, doubtless, that popular evergreen with the Victorians featured in their garden. Today we like shrubs and trees to be more colourful. Laburnum, Lilac, Almond and Cherry are all favourites with house-namers and, as the rural merges imperceptibly into the exotic, so are Mimosa, Oleander and Azalea: Rhododendron is a bit too much of a mouthful.

Exotic, too, and therefore attractive are the names of far-away places, and the places have got further away over the years (like those in the travel agents' brochures). 'Tenby' and 'Polperro' have been replaced by 'Sorento' and 'Marbella', 'Lanzarote' and 'Florida', even 'Kashmir' and 'Cotopaxi'. From literature, Waverley, Kenilworth, Ivanhoe and Marmion seem to have retained their romance for people long after they stopped reading Scott. Names from Arthurian legend like Avalon and Camelot can be found everywhere still, while the few Dickens-inspired names like Dingley Dell, Copperfield and Bleak House seem positively down-to-earth.

Professor Alan Ross who invented the 'U' and 'non-U' distinction says it is non-U to put a house-name in inverted commas or to write the number in full. He is inclined to think house-names, in town at least, inherently non-U. It is U to live at plain 5A, provided the rest of the address is OK. But there are clearly degrees of non-U-ness: if your house really was The Old Bakery it is permissible to call it such, even pleasingly unpretentious to do so. 'Ye Olde' anything is shockingly non-U as well as being unhistorical. Englishmen have never said 'ye' for 'the'; it is just a misreading of the old letter thorn (written þ) which stood for 'th'. 'Balmoral' or 'Sandringham' on a small bungalow is an absurdity deflated by 'Ahrschede' on the identical dwelling next door. 'Chez Nous' (perhaps like garden gnomes) is the inexcusable on the point of becoming chic, for such are the vagaries of fashion, but 'Shay Wen' is barely excused by its dig-in-the-ribs humour.

After the aristocracy, the most solidly U section of society has always been the landed gentry and the names of their residences are geographically descriptive of their land. Emma Woodhouse and her valetudinarian father, who was undoubtedly a gentleman, lived at Hartfield, divided from the rest of Highbury by 'its separate lawn and shrubberies *and name*', while Mr Knightley, Emma's future husband, had his seat at Donwell Abbey. The snobbish vicar's wife, Mrs Elton, was for ever talking about her relatives, wealthy through trade, with their barouche-landau, who lived at 'Maple Grove' near Bath. Jane Austen surely meant us to picture 'Maple Grove' as a slightly vulgar new Regency dwelling, smart enough by the standards of fashionable Bath, no doubt, but definitely non-U to traditional Highbury. One suspects there may have been no maple trees at 'Maple Grove' and, if Mrs E's relatives were our contemporaries, they would have doilies, serviettes for napkins, requisites in the toilet, switch on the electric logs in the grate and commit all the other solecisms so wickedly satirised by John Betjeman in *How to Get on in Society*.

Perhaps the last word in snobbish refinement is to have a substantial and expensive residence with all mod. cons and a rusticated design and call it 'Lodge', 'Cottage' or even 'Bothy'. The word 'house' itself can be snobbish when part of a name: 'Rosemead' is non-U but 'Rosemead House', Rosemead Avenue, is setting itself up above its neighbours and trying too hard to be U. A defensive or apologetic reaction is to call your house 'The Shack', 'The Pits', or 'Isore'.

It would be a duller world if, in the interests of a more classless society, we were only allowed to give our homes numbers or post codes. Homemade names lend welcome variety to otherwise monotonous streets and estates. I never had the courage to knock and see whether the residents at 'Kumincide' were welcoming or toxic, for the name suggests both possibilities, nor whether the Brooms lived at 'Genista' or the Tanners at 'Sixpenny House'; I hope so. There used to be a humble caff near Cardiff called 'Daiswill'. It must have been one of those his-and-her names unless the proprietress was hinting at services rendered beyond the expected wad and char; or could it even have stood for 'Dai swill'? I regret never finding out.

I am still wondering, too, whether 'Elveston' was a conscious anagram of 'love-nest', whether 'Lautrec' had two loos, whether it was significant that 'Pooh Corner' was near the gas works and whether the Japanese-sounding 'Fujia' might have been intended as an acronymic injunction to the unfortunate Jack. Thank goodness the house is more than 'a machine for living' - Le Corbusier's phrase. On the whole, people do not give machines names, but the entertainment value and scope for speculation provided by the names they give their houses is endless.

CAROLI LINNÆI

Naturæ Curiosorum *Dioscoridis Secundi*

SYSTEMA NATURÆ

IN QUO

NATURÆ REGNA TRIA,

SECUNDUM

CLASSES, ORDINES, GENERA, SPECIES,

SYSTEMATICE PROPONUNTUR.

Editio Secunda, Auctior.

STOCKHOLMIÆ

Apud GOTTFR. KIESEWETTER.

1740.

Systema Natura by Carl Linnaeus,
the greatast namer of them all. Title page of 1740.

Chapter 20
Miscellaneous Matters

First a little revision. Nothing is more fundamental to good gardening than dung, and the reader who remembers what was said about the 'layering' of our language in Chapter 1 may well guess correctly that 'dung' and 'dunghill' are Anglo Saxon, while 'muck' and 'midden' are Scandinavian. Old Thomas Tusser writes in 1580,

"When weather be hard
Get muck out of yard."

and he farmed in East Anglia where Scandinavian influence had been strong. 'Manure' is Norman French but didn't acquire its modern sense until the 16th century: earlier it was a verb meaning to occupy, till or cultivate land. 'Excrement', 'faeces' and 'ordure' are all examples of Renaissance 'ink-horn' terms. 'Compost' from Latin *compositus* meant a putting together or mixture and can still be used in that sense, but when we are served at table with a mixture of fruits or other foods, it is usually called a 'compote' from French to avoid unfortunate associations with the heap at the bottom of the garden. Rubbish was *robys* in the 14th century and is basically the same word as 'rubble', but their origin is obscure and the same is true of 'garbage' and 'trash'.

Gardeners

Before we go too far down that path, let us turn to gardeners who, like botanists (see Chapter 18), have surnames which can be used to illustrate the four types of British surname. These are, in the order of their frequency, local surnames, first-name surnames, occupational names and nickname surnames.

Humphrey Repton and William Kent have surnames which illustrate the first and largest of these groups. Repton is a place in Derbyshire with a famous school and Saxon church first recorded in the 8th century as Hrypadun, the hill (dune, down) of a chieftain called Hrype. Kent, from a Celtic word meaning 'borderland', was known as Cantium to Julius Caesar: it is, for some unexplained reason, much the commonest of our county surnames.

First-name surnames can be illustrated by a later pair of great contemporaries, William Robinson and Gertrude Jekyll. Jekyll comes from an old Breton name 'Iuchel' which was brought to England by the followers of Willliam the Conqueror. Robinson is obviously 'son of Robin', the commonest pet-name form of Robert, an old Germanic name favoured by the Normans and meaning 'fame-bright'.

For occupational names we could think of Percy Thrower; a thrower was probably a thread-thrower or silk-spinner rather than a potter; or of Edwin Hillier who founded the largest hardy tree and shrub nursery in the world at Winchester; a hillier made stone tiles or slates.

Finally there are the nickname surnames from appearance or character. Lancelot Brown acquired the nickname 'Capability' in his own lifetime, but his surname was also a nickname originally: most likely he had an ancestor who struck his contemporaries as notably swarthy-skinned or brown-haired. Henry Hoare (who laid out the great garden at Stourhead) had a forebear who was prematurely or conspicuously grey-haired or hoary while Christopher Lloyd who created the garden at Great Dixter has a Welsh surname which also means 'grey'. Surnames like Plant, Digweed and Pluckrose which all exist were very likely nicknames for a gardener originally, though I do not believe any famous gardeners have been blessed with such apt names: Cutbush was more likely a furze-hewer or faggot-maker.

Gardens

When we turn to the gardens these men created, their names are usually the names of the places where they are located of course. They will seldom be as obvious as Stourhead in their meaning, and guess-work rarely leads to the truth in etymology. Often they are very small places and research in the county volumes of the English Place Name Society may be necessary to track down their meaning if it is known. Early spellings are the most important evidence.

Place names cannot be quite so easily categorised as personal names though there are some useful distinctions. There are, for instance, names derived from the people who lived there, the so-called 'habitative' names. They very often end in -ham or -ton, the origins of our 'home' and 'town' but originally meaning no more than farmstead, inhabited place, at most hamlet. The first part of such names is usually a person's name, often that of an Anglo Saxon chieftain of the Dark Ages or a King's thegn who was granted the land at a later date. In either case we usually know nothing about the person except his or her name and that he or she once lived there: Hidcote (Hydica's cottage) or Petworth (Peota's enclosure) would be examples. Then there are the so-called 'topographical' names describing some feature of the landscape like Cliveden (cliff-valley), Melbourne (mill-stream) or Biddulph (by the digging or pit: the Old English verb *dulfan* also gives us 'delve').

More interestingly, place names can illustrate those 'layers' of our history and, therefore of our language which we outlined in Chapter 1. Three-quarters of all of them from the toe of Cornwall to the topknot of Scotland are Anglo-Saxon names and the four we have just mentioned are typical examples. So are Dyrham (deer-park), Saltram (home of salt-workers), Lacock (streamlet) or Wisley (meadow-clearing). However, we can find names from layers before and after the Dark Ages of the Anglo-Saxon settlement.

Celtic names are most likely to be found in Cornwall (Lanhydrock, church of St. Hydroc) or in Wales (Plas Newydd, new place) but not necessarily so: Clumber in the Nottinghamshire Dukeries is pure Celtic containing the river name *Clun* (also found in the Welsh Marches) and *bre* meaning 'hill'.

Later on, when the Danish and Norse raiders began to settle here, they added a Scandinavian layer to the names in the landscape. The areas most affected were north and east of Watling Street. Ormesby Hall at Middlesbrough would be a typical example: 'Orme' was a popular Scandinavian personal name meaning 'Serpent' or 'Dragon' ('worm' is the related English word!) and '-by', like 'thorp' and 'thwaite', is homestead or hamlet. Across the country in Cumbria we have Sizergh Castle, the *erg* (shieling or hill-pasture) of Sigrith, a Norse woman's name. Names, like plants, can be hybrids: Grimston has the Anglo-Saxon 'ton' ending joined to the Viking personal name *Gimr*, and famous Bredon Hill is composed of Celtic *bre* and Anglo-Saxon *don*, thus meaning 'Hill-Hill-Hill'.

After 1066, the names of the new Norman lords of the land were very often added to the existing place names as at Polesdon Lacey and Dunham Massey. Temple Newsam was plain *Neuhusum* (new houses) in Domesday Book but was later acquired by the Knights Templars. Montacute was known as Bishopston before the Conquest and then got the French name *Montaigu*, 'pointed hill' while no pre-Conquest name is recorded for Beaulieu, 'beautiful place'.

The Norman Conquest is not the end of the story either. New names continue to be added to the landscape but not, D.G., as the result of invasion and conquest. Claremont in Surrey got its name when the Earl of Clare bought the estate in 1714. Castle Drogo is a name not so much invented as excavated for Lutyens' great house and garden in Devon, Drogo de Teigne being the supposed ancestor of Julius Drew, the grocer whose fortune paid for the romantic extravagance. More recently still, Clough Williams Ellis devised the name Portmeirion which has now replaced the older Welsh name Aber Ia on the map.

Garden Design

For our vocabulary of garden design and ornament we have borrowed heavily from the Continent. This is partly because the taste of a whole generation of Royalist aristocrats was informed by their exile during the Commonwealth period, and then, in the following century, the Grand Tour became a standard part of the education of a young gentleman. The preference for landscape which swept away much of the formality of earlier centuries was also reinforced if not directly inspired by the paintings of Claude Lorraine which became immensely popular. Even the word 'landscape' which looks homegrown is in fact a Dutch painting term, *landschap*, while the exotic-looking 'gazebo' is an exception in probably being an

English invention (see below). 'Patio' is Spanish, 'obelisk' Greek and we have borrowed 'barbecue' from Spanish America quite recently but the great majority are from Italian or French including bust, statue, urn, canal, column and the following which I give alphabetically with the dates when they were first used in England according to the Oxford English Dictionary.

Arbour from Norman French was originally *herber*, a herb or vegetable garden but had already become a bower (the English word) or shady retreat in Chaucer's verse by about 1385.

Avenue comes from French *avenir*, the prospect before one. It is used in Evelyn's Diary in 1645 and at that time meant the chief approach to a country house, often but not necessarily bordered by trees.

Garage was borrowed from the French *garage pour autos* in 1902 and comes from an Old French verb meaning to care for. In France it retains a wider sense than in England being used of docks, canal basins, boat houses (*garage pour canots*), railway sidings and so on.

Gazebo seems to have appeared first about 1752 for a belvedere or look-out place originally on the roof of a house. Early on, garden ones were often designed in the Chinese style so it is possible that the name is a corruption of an oriental word but most authorities think it a jocular formation from the verb to gaze in imitation of Latin future tenses like *videbo*, 'I will see'.

Grotto comes ultimately from Latin *crypta*, a vault, but appears in the Italian of Dante as *grotta* or *grotto*. It comes into English in 1617 for a small cave but in the following century artificially constructed ones ornamented with minerals, fossils, shells, etc. became popular features of the landscape garden. The poet Pope built a famous one at his villa at Twickenham.

Ha-Ha first appears in England in 1712 as 'Ah-Ah' in a translation of a French work on gardening. The sunken wall allowing an uninterrupted view from garden out across the landscaped park or countryside became a crucial feature of what became known as the English style of gardening. Ha-Ha is plausibly thought to come from one's exclamation of surprise at first coming upon an ingenious hidden obstacle.

Lawn from Old French *launde* was originally a glade in woodland or uncultivated grassy ground. It only acquired the modern sense of a level area of grass kept mown in the 18th century and appears in Miller's Gardeners' Dictionary, 1733. 'Lawn' as the name of a fine linen is unrelated, coming from the town of Laon in France.

Loggia in Italian is an open gallery or arcade giving shelter from the sun but in England more often from wind or rain. The words 'lobby' and 'lodge' are related and the dictionary's earliest example is from Defoe's *Tour of Great Britain*, 1742.

Parterre in modern French means a flower bed (or the pit of a theatre) but we borrowed the word in the 17th century - the poet Carew uses it in 1639 - for a level area with ornamental flowerbeds, not unlike the Tudor knot garden, but often on a much grander scale *par terre*, across the ground.

Patio first appears in English in 1808 used for a paved area adjoining a house though in Spanish it is usually an open inner courtyard.

Pergola from Latin *pergula* by way of Italian was a projecting roof often used as a vine arbour. By the time John Evelyn uses the word in 1675 it clearly has a removable top: 'Twixt East and North erect a *Pergola* or shed, so contrived with a Cover, as to exclude or admit the weather at pleasure'.

Terrace comes by way of Old French from Latin *terra* and came to mean a raised level walk. This was the vantage point from which one could appreciate the design of the parterre. It appears in English in 1575.

Topiary arrived in 1592 from French *topiare*. Rabelais uses the word in 1542. In Latin *topia* is ornamental gardening and that is the sense in which the word was first used here before becoming restricted to the fashionable practice of clipping and shaping shrubs.

Trellis comes from medieval French *treliz*, a coarsely woven fabric. It is being used for lattice (another French word) erected as a support for fruit trees or climbing plants as early as 1513.

Gardeners and their families are not the only creatures to frequent gardens. We have already given birds and their names an airing in Ch.17 and an estimated six million domesticated cats must have some impact on their numbers, but what of wild mammals? In some places even the largest of them, the deer, can be devastating visitors who deprive us of the excitement of seeing them by coming and going while we are still in bed. 'Deer' comes from a root word which meant 'breathing creature' and could be applied to any animal, only later becoming 'particularised': Edgar acting the part of Mad Tom in King Lear sings of 'mice and rats and such small deer'. 'Fallow', incidentally, is an Old English word for reddish-yellow which only survives in the English name for *Dama dama*, our commonest deer: it has nothing to do with uncropped arable land.

Other creatures like foxes and badgers also make nocturnal visits and, as modern farming makes the countryside less hospitable to them, so they have become more familiar to the urban gardener. 'Fox' (*Vulpes* in Latin) is a Germanic word which goes back to a root meaning 'the tailed one' and early on became a synonym for a cunning thief. I have seen a fox eating blackberries and cherries and he may raid your dustbin but does not do the gardener much harm unless he keeps poultry. The badger will grub up roots and overturn a beehive to eat the contents, but does most of us no mischief. His Latin name *Meles* looks as if it ought to be connected with *mel, mellis*, honey. He was originally a 'badgard' probably because of the white badge on his head, but there are other theories.

If you find a hedgehog's skin with the contents neatly scooped out, that is probably the badger's work as he cares as little for prickles as for bee stings. 'Hedgehog' needs no explanation: it is *erinaceus* in Latin, or *echinus* which crops up in the scientific name of several prickly plants and in *echinoderm*, (hedgehog skin), the class of creatures containing sea urchins.

Peter Rabbit is all too well known to the Mr Macgregors of this world, though less numerous since myxomatosis than formerly, but the most serious damage he does is barking young trees when snow is on the ground rather than over-eating himself on lettuce and carrots. 'Rabbit' is thought to be diminutive of Flemish *robbe* but it goes back no further in English than the 14th century. The Normans are known to have introduced (or reintroduced) rabbits to England and the mediaeval name was the Old French *coney* from Latin, the scientific name being *Orcytlagus* (from Greek 'digging-hare') *cuniculus*.

The house mouse, *Mus musculus*, (little mouse) can be found anywhere man lives from arctic to the equator and has probably been in Britain since the New Stone Age but is not strictly native. 'Mouse' which is obviously cognate with the Latin *mus* comes by the Germanic route into Old English. The root meaning in Sanskrit is 'robber'. The harvest mouse, *Micromys minutus*, and *Apodemus flavicolis*, the very local yellow-necked mouse are not really garden species but the woodmouse or long-tailed field mouse, *Apodemus sylvaticus* certainly is: the names mean 'away from home' and 'of the woods' but, if they can get into your garden shed, they will enjoy your stored apples or potatoes and are canny at finding peas and beans as soon as you sow them. They are much more red-brown than house mice, and white underneath.

Voles have blunter snouts and smaller ears than mice but still seem to confuse many people. They were originally called vole-mice from Old Norse *vollr*, a field, our word 'wold' in fact. The field vole, *Microtus* (small-eared) *agrestis* (of the fields) is often wrongly called the 'short-tailed field mouse'. This is the Timmy Willie of Beatrix Potter's *Tale of Johnny Town Mouse*. The watervole is not a rat, but Kenneth Grahame was only following the common error when he called his character 'Ratty' in *Wind in the Willows*. The bank vole which has longer ears and tail than the field vole and prefers less open country is the one most at home in the garden. Its name *Clethrionomys glareolus* raises several problems but probably means something like 'bolt-toothed mouse of gravelly places' though it actually prefers earth banks.

Of our native shrews two, the common *Sorex araneus*, and the smaller pigmy shrew *S. minutus*, like rough grass and hedgerows and make their presence known to the gardener since they are active by day and do a great deal of snuffling and squeaking as they search for insects and worms. 'Shrew' is related to an old Germanic word for 'dwarf' but their distinguishing features are their very slender, pointed snouts. They were once believed to be venomous and that is reflected in the name of the common shrew, *S. araneus* meaning 'of a spider' which also had that reputation. It probably also accounts for the fact that, since the Middle Ages, an ill-natured person (originally a man rather than a woman) has been called a shrew.

Gardeners only have one species of mole to contend with, *Talpa europea*, which is just as well since they can be a great nuisance. I knew a gentle elderly philosopher who was turned to apoplectic rage by their depredations in the garden which was his pride and joy. 'Mole' is an abbreviation of 'mouldwarp', earth-thrower.

In most parts of Britain now there is only one squirrel too, the native red *Sciurus vulgaris* having been replaced by the larger N. American grey *S. carolinensis*, which was introduced in 1890. The grey are the more destructive of trees and it may have been foresters who put it about that they are 'nothing but tree-rats': in fact the two species are closely related, the red preferring coniferous and the grey broad-leaved trees. It is also a calumny that the grey killed off or drove out the red, though it may have prevented them from reoccupying their former territory after they were almost exterminated by disease. 'Squirrel' comes via French and Latin from Greek *skiuros* meaning 'shade-tail'.

Stoat and weasel may be occasional visitors to country gardens: both are Old English names whose etymology is obscure. Bats, however, have a name abbreviated from the excellent Scandinavian *lethrblaka*, 'leather-flapper', which had become *bakke* in Middle English. The Old English name was 'reremouse' from a verb indicating rapid movement while 'flittermouse', a country name in some parts, is identical to German *fledermaus*. We have twelve of these, the only true flying mammals, the commonest being the smallest of them. 'Pipistrelle' was adapted from the French by the 18th century naturalist Thomas Pannant and comes from the Latin *vespertilio* (*vesper* = evening). He also borrowed the name 'noctule' from the

French of Buffon for our largest bat which has a 14-inch wingspan: it was a less happy choice of name since this species is the one most likely to be seen flying in daylight.

If you have frogs in your pond, they are likely to be *Rana temporaria*; the specific name implies the 'adaptability' of their amphibious behaviour; though other species like the edible frog, *R. esculenta*, are established in some places in the south of England. The French call them *grenouille* and get called 'frogs' themselves (which is the name in the Germanic languages) because of their gustatory predilection. A toad was *tada* in Old English and still regarded as venomous in Shakespeare's time. 'Tadpole' comes from this source, literally 'toad-head (poll)' and a newt was originally 'an ewet': 'asker' is its common dialect name.

Creepy-Crawlies

When we descend to creatures without backbones, the invertebrates, the numbers which inhabit the garden become astronomical. Many are invisible to the naked eye, most do not have English names and we shall have to confine our attention to some few of these which are most likely to have come to the attention of the gardener.

Earthworms are not too much of a problem: Old English *wyrm* also meant 'serpent' or 'dragon'. They belong to the phylum *Annelida* (ringed or segmented worms) and to the group which is called *Oligochaeta* which means 'few bristles' to distinguish it from the marine bristle-worms. The few bristles they do have which you can see if you take the trouble are backward-pointing and on the base of the segments. They enable the worm to move through the soil, help to anchor it in its burrow and mean that a bird must swallow it head first. There may be three million of them under an acre of old grassland but there are only 25 species in this country, ten of them common, and only two of which make worm-casts on the lawn, *Allolobophora longa* and *nocturna*. Anglers, I believe, favour the red worm, *Lumbricus rubellus*, rather than the larger, commoner *L. terristris* and the gardener may even be able to give a name to the dark red worm which multiplies prodigiously in his compost heap and also lives under the bark of dead wood: it is called the brandling worm because it is marked or branded with a conspicuous orange band or 'clitellum' which is actually the mature worm's egg-laying organ and not, as is sometimes thought, a scar where it has been chopped in half and joined together again (something worms cannot do though the head end can grow a new tail). The gardener may also have seen clusters of small greyish-white worms in his compost: they are known as pot-worms but whether because they frequent potting compost or wriggle to and fro into the shape of pothooks or for some other reason, I have been unable to determine.

So far so good, but there are round worms as well as segmented ones, the nematodes, a name which means 'thread-like'. There may be 20 million of them to the square yard, mostly microscopic, and the gardener as well as his soil has a large population of them thriving within and upon him. Oddest of all is the thunder worm, like a piece of brown cotton over a foot long which sometimes climbs over wet foliage during summer showers. Most troublesome are vegetable-eating nematodes like the potato eelworm: others are carnivorous as are the black flatworms which mostly live in ponds.

Molluscs are soft-bodied (Latin *mollis*) creatures. There are a great many of them, including squid and octopus, but the only two groups of concern to the gardener, the snails and slugs, are both *gastropods* (stomach-footed). If you have many snails in your garden (there are about twenty common species) the soil must be alkaline since they need lime to make their shells. The name was *snaegel* in Old English and snail's-pace in German today is *Schneckengang*. Slugs are shell-less snails which are all too happy on my acid soil and the name, which is also used of a lead bullet and a line of printer's type, is Scandinavian in origin for something heavy, slow-moving or indolent. Most of the dozen species which are common in gardens eat decaying vegetable matter or fungi for preference though some are carnivorous. The wise gardener will single out for his special wrath the grey or whitish netted slug which eats green leaves and has a great liking for lettuce and the garden slug which is smallish, dark-coloured with a yellowish or orange sole which does most damage to roots and emerging shoots.

All the remaining invertebrates in the garden which we need to mention are *Arthropods* which means literally 'jointed foot': they all shed their skin as they grow, have bodies divided into distinct segments, and jointed legs, though they are often of most concern to the gardener in their legless larval stage. Wireworms and cutworms are such larvae and not worms at all. Ditto the so-called bloodworm which wriggles about in the water butt, a gnat larva, and the glow-worm which is a female snail-eating wingless beetle.

Woodlice are *crustaceans* (Latin *crusta*, a shell or crust) and are related to crabs and lobsters. Because they were originally aquatic, they will quickly dry up and die if they stray from moisture for more than a short time. As their name suggests, they live on rotting wood and other decaying vegetable matter but will eat seedlings in a greenhouse. Over 40 species have been recorded in Britain but only six are common in gardens. The pillbug which rolls up into a ball is *Armadillidium vulgare* and is not only swallowed by schoolboys for a dare, but was prescribed by doctors for digestive ailments and given to cattle, hence the name 'cudworm'. There are many other country names like sowbug, slater, coffin-cutter and tiggy-hog.

Centipedes and millipedes are sometimes lumped together as *Myriapodes* (many feet) though they are not particularly closely related. There are about 40 British species of each and, considering how common they are, it is surprising that, unlike the woodlice, they do not seem to have any common English names. Centipedes may have from 30 to 202 legs, one pair to each section of the body. They are mainly carnivorous with poisonous fangs for killing their prey which includes worms, woodlice and spiders, and large tropical species can give humans a dangerous bite. Millipedes are mostly black and vegetarian, with stink glands to repel enemies and some species curl up like a watch spring when disturbed. They generally have two pairs of legs to each body segment and as the maximum number of segments possessed by a British species is 60, the maximum number of legs they can have is 240, less than a quarter of the number their name implies.

Arachnophobes, like Miss Patience Muffet who was the

daughter of a 17th century entomologist, had better skip this paragraph. *Arachne* is the Greek for spider and the class to which they belong together with ticks, mites, harvestmen and scorpions is the *Arachnida*. They are unlike the insects in having no nymph or larval stage, four rather than three pairs of legs and, in the case of spiders, the head and thorax combined so that their bodies have two rather than three sections. A 'mite' was a coin of very little value as well as a tiny creature which may be found thriving almost anywhere including ripe cheese and old books. The scarlet velvet mite that often swarms over paving stones in the summer is carnivorous and does the gardener no harm. His two main enemies are the gall mite which causes 'big bud' on blackcurrants and red spider mite which only became a serious pest on fruit trees after DDT killed off most of its natural predators; reintroducing one or more of them is the best control in a peach house.

When we come to the spiders, of which there are nearly 600 British species and sometimes an estimated two million to an acre, none can be regarded as an enemy of the gardener; they are all carnivorous and eat harmful and beneficial insects indiscriminately. 'Spider' was *spithra* in Old English and connected at some stage in its history with 'spinner'. The so-called garden spider (also known as the diadem or cross spider because of the pattern of white dots on its abdomen) is typical of the spiders which spin beautiful orb webs. The species which get into our houses and garden sheds are more likely to make sheet webs with a tunnel leading to a safe retreat in a crevice. It is some members of the *Tegenaria* genus which cause most alarm as they can be large, black, hairy and scuttle across the floor very fast. Spiders have fangs with which they poison or paralyse their prey but no British species can give a human being a significant bite. The non-web-making spiders include jumping spiders like the little black and white zebra spider which pounces on its prey, the hairy wolf spiders which outrun theirs and the crab spiders which lie in wait for insects often in an open flower. There is even a spitting spider which can fire darts of sticky thread to pin down its quarry.

Finally we come to the largest class of *arthropods*, the insects, with at least a million species world-wide and 20,000 British ones, most of which will have turned up in somebody's garden at some time. 'Insect' comes from Latin *insecare*, to cut up into sections, and they all have bodies divided into *three* parts: head, thorax (with *three* pairs of legs) and abdomen. They are the only *arthropods* to have wings (though not every one has them) and the thirty orders into which the insects are classified are mostly named from the type, arrangement or number of their wings. For example, the *Lepidoptera* (scale-wings) are the moths and butterflies, the *Coleoptera* (sheath-wings) are the beetles, the *Hymenoptera* (membrane-wings) include bees, wasps and ants and the *Diptera* (two-wings) are the flies.

Most insects are fairly small, the largest in this country being the stag beetle, the great green bush cricket and some of the hawk-moths. The old rhyme about great fleas having lesser fleas upon their backs to bite 'em is not far wrong: take a close look at a bumble bee, for example, and you will very likely see tiny mites scurrying about its hairy body and many insect species, notably the fleas of course, are similarly parasitic. Many insects, too, go through amazing transformations in their lives by way of a pupa or chrysalis, from 'woolly-bear' caterpillar to garden tiger moth, for example, or from white maggot to bluebottle or blowfly. Flies fly, tiger-moths are striped, bluebottles are blue and most of the English names of garden insects like grasshopper, hover-fly, lacewing, leather-jacket, orange-tip, shield-bug are similarly straightforward and descriptive. Below, alphabetically rather than by their scientific orders, are notes on some of the less obvious ones.

Ant is a shortened form of earlier 'ammet' or 'emmet', which is still a common country name, exists as a nickname-surname and crops up in placenames like Ampthill, Beds. (earlier 'Amethyll'). 'Pissmire' is another country name from the urinous smell of anthills.

Aphis was the name adopted by Linnaeus for a plant-sucking louse. The name was used by earlier entomologists but originally seems to have been a misreading of a Greek word for a bug. 'Aphid' is really a mistaken 'back-formation' from the correct plural 'aphides'. Plant sap is rich in sugars but poor as a source of protein so sap-sucking insects get an excess of sugar in their diets which they excrete as droplets of sticky 'honeydew': this attracts ants and also sooty mould.

Bee goes back to Old English *beo* and ultimately to a root meaning of tremble or quiver.

Beetle originally meant 'biter' and they are, indeed, biting rather than sucking insects. They often have conspicuous mandibles which, however, can do no harm to human skin, unlike the hypodermic needles of some of the suckers.

Butterfly is not a Spoonerism for 'flutter-by' as Lewis Carroll would have you believe, but seems to mean what it says. There are names like 'milk-thief' in other languages reflecting a belief that the insects are attracted to dairies. The fact that one of the earliest and most conspicuous butterflies to appear each year is the butter-yellow brimstone may have helped the name to stick. Least popular with gardeners are the so-called 'cabbage-whites': there are three species, the large, small and green-veined whites which may all lay eggs on brassica plants. Another small white butterfly is the female orange-tip which is looking for cruciferous weeds and not cabbages.

Capsid Bugs get their name from Latin *capsa*, a box, from the rectangular shape of their abdomens. There is a black-kneed one which usefully eats red spider-mites but the common green capsid damages apples often causing brown scabs on the fruit.

Caterpillar. Some ingenious stories have been made up to account for the odd-looking word but in fact it comes from Old French *chatepelose*, 'hairy-cat'. The modern French name *chenille*, however, comes from Latin *canicula*, 'little dog'.

Chafer. The garden chafer with green head and thorax is a small relative of the maybug or cockchafer. Both adult beetles and their grubs chew or gnaw - that is what the word means - and can be a nuisance if too numerous. The verb to chafe meaning to warm, inflame, fret or rage is not related.

Cockroach is from Spanish *cucaracha* but the etymology is unknown.

Cricket was *criquet* in Old French and is imitative of the male insect's stridulation which is aimed at attracting the female. In its other sense, a cricket may have originally been the bat or the wicket used in playing ball games before it became the name of a specific game: *krick* is a stick in Flemish.

Earwigs are nocturnal insects nibbling your Dahlias in the dark and then seeking a hole like the end of a cane or an ear to hide in by day. They are 'ear-worms' in several other languages and it must once have been a common experience to be woken up by an unwelcome visitor.

Fleas are blood-suckers of course so the gardener is most likely to encounter them on his pets, in an old bird's nest or when disentangling a hedgehog from his netting. The word is unconnected with flebotomy or phlebotomy, the old medical practice of blood-letting, but comes via Old English from a Germanic root, possibly the same as to flee or hop off in a hurry.

Froghopper is an obvious enough name but they are also known as spittlebugs because the young nymphs exude the familiar frothy 'cuckoo-spit' to protect themselves from both desiccation and predators.

Gnats spend most of their lives as larvae inside little galls on plants like swede and Germander Speedwell. The name has been in the language since Anglo Saxon times.

Hornets, like all members of the wasp family, are not aggressive unless disturbed. The name probably originally meant 'horn-bearer' but whether in reference to the formidable sting or the horny brown appearance of the thorax is uncertain.

Ladybirds and their larvae eat many times their own weight of aphids and are therefore the best known and loved of all the beetles. 'Ladybird' was a term of endearment in Shakespeare's England and applied to the beetle as early as the 17th century when they were also known as Ladybugs or Cowladies. It is presumed to be a contraction of 'Our Lady's bird' perhaps because they first emerge from hibernation about Lady Day (March 25th). Commonest is the largish red 7-spot ladybird but there are also 2-, 10-, 22- and 24-spot species: the last of these must have one of the longest of all scientific names, *Subcoccinella vigintiquatuorpunctata*.

Loopers are the caterpillars of the *Geometer* or 'earth-measurer' moths. Only having legs at the front and the rear, they move by bringing the hind legs up to the front ones making a loop of their bodies as they progress.

Midge has been in the language since before the Conquest though 'midget' for a very small person only goes back to 1865. It is ultimately related to Latin *musca*, a fly.

Mosquito also comes from Latin *musca*. It was borrowed by Elizabethan voyagers like Hakluyt from the Spanish or Portuguese where it means 'little fly'. One of the commonest species which bites man is the well-named *Culex molestus*.

Moths are distinguished from butterflies by not having club-tipped antennae and being mainly night-flying. There are also a great many more of them, about 2000 British species compared with fewer than 60 butterflies. 'Moth' goes back to Old English and there are related names in other Germanic languages. Least popular with gardeners are the cabbage moth whose well-camouflaged caterpillar goes straight to the heart of the cabbage and the codlin moth whose pinkish grubs burrow into apples.

Sawflies are so called because the female has a saw-like ovipositor with which she lays eggs in leaves and twigs where they often cause galls. Their larvae may be a nuisance on gooseberries, apples, roses and many other plants.

Thrips is both singular and plural: they are slender black insects most often seen crawling among the stamens of flowers which they help to pollinate. Too many are bad news as they suck sap and may prevent fruit forming. In sultry weather they often take to the air in huge numbers and so get the name of thunderbugs. 'Thrips' comes from the Latin of Pliny from the Greek for a woodworm.

Wasp. 'Beware of wappys' says a 15th century writer and the dialect 'wops' is the older form in English and to that extent the more correct name. It comes, however, from Latin *vespa* and is connected with words like 'web' and 'weave' because of the construction of the wasp's nest.

Weevil is an older name than beetle and in Old English the glow-worm was *goldwifel* and the dung beetle *tordwifel*. Today we use it for slender-snouted beetles, some of which like the vine, pea and raspberry weevils can be garden pests.

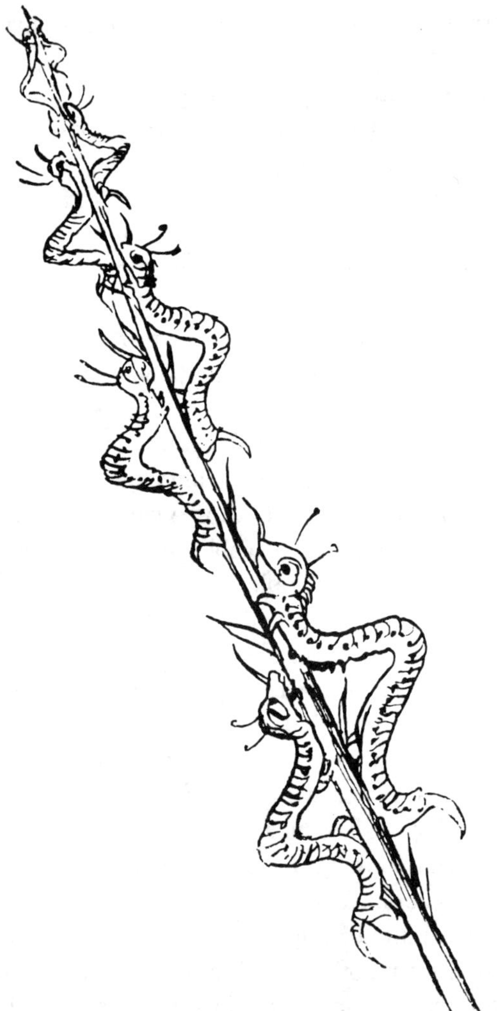

Nasticreechia krorluppia
from Edward Lear's *Nonsense Botany*, 1871.

Bibliography

Brown, Jane. *Eminent Gardeners* Viking 1990

Carter, H. Gilbert *Glossary of the British Flora* C.U.P. 3rdEd. 1964

Coats, Alice M. *Flowers and their Histories* A & C Black, 2nd Ed 1968

Coombes, Allen J. *The Collingridge Dictionary of Plant Names* Collingridge 1985

Chinery, M *The Natural History of the Garden* Collins 1977

Chinery, M. *The Collins Guide to the Insects of Britain & Western Europe* Collins 1986

Desmond, Roy *Dictionary of British & Irish Botanists and Horticulturalists* Taylor & Francis 1977

Gledhill, D. *The Names of Plants* C.U.P. 2nd Ed.1989

Gotch, F.A. *Mammals: Their Latin Names Explained* Blandford 1979

Griffiths, Mark *Index of Garden Plants* R.H.S & Macmillan 1994

Grigson, G. *The Englishman's Flora* Phoenix House 1958

Grigson, G. *A Dictionary of Plant Names* Alan Lane 1974

Hadfield, M. et al *British Gardeners: A Biographical Dictionary* A Zwemmer 1980

Hyams, Edward *Plants in the Service of Man* J.M. Dent 1971

Hill, Thomas A. *The Biology of Weeds* Edward Arnold 1977

Jackson, Benjamin *A Glossary of Botanical Terms* Duckworth 1949

Jaeger, E.C. *A Source Book of Botanical Names & Terms* Chas. C. Thomas Illinois

Johnson A.T.& Smith A.H. *Plant Names Simplified* Landsman's Bookshop 1972

Lockwood, W.B. *The Oxford Book of British Bird Names* O.U.P. 1984

Potter, S & Sargent, L. *Pedigree: Words from Nature* Collins 1961

Prior, R.C.A *On the Popular Names of British Plants* Williams & Norgate 1870

Salisbury, Sir E. *Weeds as Aliens* Collins 1961

Savory, T.H. *Naming the Living World* Edinburgh U.P. 1962

Savory, T.H. *Latin and Greek for Biologists* Merrow 1971

Smith, A.W. *A Gardener's Dictionary of Plant Names* Cassell 1972

Stearn, W.T *Botanical Latin* David & Charles 1983

Stearn, W.T. *Stearn's Dictionary of Plant Names* Cassell 1992

Index

The meaning of a specific name is explained under the entry for the generic name. Where the English and scientific name of a plant are nearly identical (e.g. Borage/Borago, Rose/Rosa, Tamarisk/Tamarix), one or other is included but not both.

Aaron 17
Abutilon 93
Acaena 109-10
Acalypha 78
Acantholimon 118
Acanthus 64
Acer 93
Achillea 64
Achmea 78,80
Acidanthera 128
Aconitum 64,69
Acorus 21
Acuba 91
Adam's Flannel 76
Adam's Needle 76
Adiantum 78
Adonis 64
Adoxa 25
Aegopodium 145
Agropyron 145
Aethusa 143
African Daisy 47
African Lily 65
African Violet 43,88
Agapanthus 65
Agaricus 161
Ageratum 46
Aglaonema 78
Agrostemma 46
Ajuga 110
Alcea 46
Alchimella 110
Alkanet 65
Allamanda 78
Alliaria 30
Allium 6, 128, 161, 163,170-1
Almond 176
Aloe 78
Alonsoa 47
Alpine Lady's Mantle 28
Althea 46-7
Aluminium Plant 87
Alum Root 39
Alyssum 118
Amaranthus 47
Amaryllis 85,128,132
Amazon Lily 77
Amalanchier 93
American Cowslip 120
Amer. Wych-Hazel 38
Ammobium 47
Anacyclus 118
Anagalis 119
Ananas 78,80
Anaphalis 65
Anchusa 65

Anethum 170
Angel's Fishing Rod 130
Angel's Trumpet 83
Angel's Wing's 80
Anemone 9,10,64,128
Angelica 168
Anise 168
Aniseed 168
Annual Meadow Grass 142
Ant 224
Antenaria 119
Anthemis 5
Anthionem 118
Anthoxantheum 21
Anthriscus 170
Anthurium 78-9
Antirrhinum 45,47
Aphelandra 79
Aphis 224
Apium 157
Apple 20,176
Apple of Peru 57
Apricot 177
Aquilegia 26
Arabian Primrose 65
Arabian Violet 84
Arabis 119
Arbour 215
Arctotis 47,60
Arenaria 119
Argemone 48
Arisarum 110
Aristolochia 103
Armeria 119
Armorica 2,171
Arnebia 65
Artemisia 14,22,173
Artichoke 154
Artillery Plant 87
Artiplex 48
Arum 4,17,44,129-30
Aruncus 65
Asarum 110
Asclepias 66
Ash 20
Asparagus 7,64,154
Asperula 21, 111
Aspidistra 79
Asplenium 79
Ass-Catcher 58
Aster 14,66
Astilbe 66
Astrantia 66
Aubergine 153
Aubretia 36
Auricula 124
Autumn Crocus 29,129
Avenue 215
Avocado 153
Azalea 93-4

Baby Blue-eyes 57
Baby's Breath 70
Bacc(h)ar 111
Badger 217
Balloon Flower 74
Balm 168
Balsam 53,168

Banks, Sir Joseph 199
Banksia 35
Barbados Lily 85,132
Barberry 94
Barrenwort 112
Bartonia 55
Basil 169
Bastard Dittany 69
Bat 219
Bath Asparagus 135
Bay Tree 92,169
Bean 154
Beard Tongue 73
Bear's Breeches 64
Beaulieu 214
Beauty Bush 40
Bedstraw 21,22,26
Bee 225
Bee Balm 41
Bee Bread 169
Bee Catcher 51
Beech 20
Beefsteak Geranium 36
Beefsteak Plant 58,78
Bee Orchid 25
Beetle 225
Beetroot 153,155
Beggar's Blanket 60
Begonia 36
Bellis 147
Bells of Ireland 56
Beloperone 79
Bentham, George 199
Berberis 94
Bergamot 41,169
Bergenia 111
Beta 153
Betony 114
Bible-leaf 7
Biddulph 213
Biddy Bid 109
Bignonia 35,103
Bilberry 177
Billbergia 80
Bindweed 102,145
Bird Cherry 178
Bird-of-Paradise Flower 43
Bird's-foot Trefoil 26
Bird's Nest Fern 79
Bird's Nest Orchid 26
Birthwort 103
Bishop's Hat 112
Bitter Cress 30
Blackberry 177
Blackbird 186
Blackcap 185
Black-eyed Susan 42,106
Black Locust Tree 16
Blanket Flower 38,76
Blazing Star 55,71
Bleeding Heart 68
Blister Plant 121
Blood Flower 66
Bloodroot 125
Bloody Burnet 74
Bluebell 18-19

Blueberry 177
Blue-eyed Mary 113
Blue Jacket 74
Blue Lace Flower 60
Blue Marigold 66
Bluet 122
Blue Tit 195
Boddle 140
Bog Asphodel 28
Bog Violet 26
Bone-Set 114
Borage 65,169
Bottle 141
Bourgainvillea 35,103
Box 91,94
Brachycome 48
Bramble 6,177
Bramley 20,176-7
Branke Ursine 64
Brassica 153,156,171
Bredon Hill 214
Broad Bean 154
Broccoli 155
Brodiaea 129
Bromeliad 80
Broom 20,26,96
Broomrape 26
Broom Toad Flax 53
Browallia 36
Brown, Lancelot 212
Brunfels Otto 56
Brunfelsia 36
Brunnera 111
Brussels Sprout 27
Buddleia 33-4,36
Bugle 110
Bulbous Violet 131
Bullfinch 186
Burnet 6
Burnet Saxifrage 119
Burning Bush 53,69
Busy Lizzie 53
Butterbur 151
Buttercup 17,148
Butterfly 224-5
Butterfly Bush 34
Butterfly Flower 59
Butterwort 26,124
Buxus 94
Cabbage 156
Cabbage Palm 82
Calabrese 156
Caladium 80
Calandria 119
Calceolaria 119
Calendula 17,48
Calico Bush 16,40
Californian Poppy 37
Calla Lily 35,44
Callirohe 120
Callistephus 48
Calluna 94
Calochortus 129
Caltha 17
Calve's Snout 47
Calystegia 145

Camassia 129
Camellia 36
Campanula 18,27
Campion 61,72,126
Campsis 103
Canadian Fleabane 28,69
Canary Creeper 102,106
Candleberry 149
Candlemas Bells 131
Candy Mustard 118
Candytuft 48,112
Canterbury Bells 27
Cape Cowslip 85,133
Cape Gooseberry 57,64,178
Cape Jasmine 38
Cape Primrose 88
Caper Spurge 150
Capsella 142
Capsid Bug 225
Caraway 169
Cardamine 143
Cardiocrinum 129
Cardoon 156
Carduelis 185,188
Cardus 185
Carnation 68
Carpenteria 36
Carpenter's Apron 23
Carrot 156
Carum 169
Caryopteris 94
Cassava 76
Cassiope 120
Castanea 178
Cast Iron Plant 79
Castle Drogo 214
Castor Oil Plant 59
Catanche 66
Catchfly 126
Caterpillar 225
Cathedral Bells 104
Catmint 113
Cattleya 81
Cauliflower 157
Cautleya 66
Ceanothus 94
Celandine 19,147
Celery 157
Celeriac 157
Celosia 49
Centaurea 67
Centipede 222
Centranthus 49,67
Cerastium 111
Ceratostigma 81
Chaenomeles 15, 94, 103
Chafer 225
Chaffinch 187
Chamaedorea 81,86
Chamaemelum 31
Chamaerops 81,86
Chamomile 31
Chard 157
Charity 74
Checkerberry 112
Cheddar Pink 27

Cheiranthus 49
Chenille Plant 78
Cherry 178
Cherry Laurel 93
Chervil 170
Chestnut 178
Chevisaunce 49
Chickweed 142
Chicory 157
Chilean Bellflower 57,102,104
Chilean Crocus 137
Chilean Fire Bush 95
Chilean Glory Flower 104
Chilean Potato-Tree 106
China Aster 48
Chinchinchee 135
Chinese Evergreen 78
Chinodoxa 129
Chives 64,170
Chlorophytum 81
Choisya 36
Christmas Cactus 35,43,88
Christmas Orchid 81
Chyrsanthemum14,20,54,67 74-7
Chrysogonum 67,111
Cigar Flower 50
Cineraria 46,49
Cinquefoil 25
Cirsium 146
Cissus 81
Cistus 94
Citrus 81
Clapham, A. R. 203
Claremont 214
Clarkia 14,36-7,52
Cleavers 27,101
Clematis 14,35,102-3
Cleome 49
Clerodendron 81
Clianthus 103
Cliveden 213
Clivia 36
Clover 24-5
Clumber 213
Coal Tit 195
Cobaea 102,104
Cob-nut 178-9
Cochlearia 22
Cockroach 225
Cock's Comb 49
Coconut 31
Codiaeum 82
Codlin 7
Codonopsis 120
Colchicum 14,129-30
Coleus 82
Colewort 156
Collared Dove 187-8
Collinsia 50
Coltsfoot 25
Columba 197
Columbine 26
Comfrey 114
Cone-flower 42,69
Convallaria 111
Convolvulus 50

Cool Tankard 169
Copperleaf 78
Coral Flower 39
Coral-Root 26
Cordyline 82
Coreopsis 59,67
Coriander 170
Corn 158
Corn Cockle 46,140
Cornel 94
Corn Flag 132
Corn Flower 69,141
Corn Marigold 67,140
Corn Salad 21,67,158
Corn Poppy 46,141
Cornus 94-5
Corvus 190
Corydalis 129
Corylus 95,178
Cosmos 50
Cotinus 95
Cotoneaster 95
Cotton Lavender 99
Cotula 50
Couch-Grass 145
Cow Parsnip 24
Cowslip 28-9 124
Crab Apple 179
Cranberry 28
Cranesbill 70
Crassula 82
Cream Cups 58
Creeping Burnet 109
Creeping Jenny 19
Creeping Wintergreen 112
Creeping Zinnia 125
Crepis 120
Cress 21,158
Cricket 226
Crimson Glory Vine 107
Crocosmia 15,41,130
Crocus 13,130,172
Cross Vine 103
Croton 82
Crown Daisy 67
Crown Imperial 131
Cryptanthus 80
Cuban Lily 136
Cuckoo-Flower 29
Cuckoo-pint 5
Cucumber 159
Cucumis 181
Cucurbita 160,162
Cuddy-Lugs 60
Culpepper, Nicholas 201
Cup-and-Saucer Plant 104
Cup Flower 123
Cuphea 50
Cupid's Dart 66
Currant 179
Cyclamen 130
Cydonia 14,15,183
Cymbidium 82
Cynanthus 120
Cynara 154,156

Cyperus 82-3
Cypripedium 83
Cytisus 95
Dahlia 37
Daffodil 29,134
Daisy 27,147
Daisy Bush 41
Damson 179-80
Dandelion 25,146-7
Danewort 3
Daphne 92
Darwin, Charles 202
Datura 83
Daucus 156
Day Lily 71
Deadly Nightshade 51,57
Deadman's Bells 51
Dead Nettle 144
Deer 217
Delphinum 45,68
Dendrobium 83
Deutzia 37
Devil-in-a-Bush 57
Devil'd-Bit Scabious 26
Devil's Claw 123
Devil's Fig 48
Devil's Ivy 88
Dianthus 68
Diascia 50
Dicentra 68-9
Dictamnus 69
Didiscus 51,60
Dieffenbachia 83
Dierama 130
Digitalis 51
Dill 170
Dimorphotheca 51,58
Dizygotheca 83
Dodecatheon 120
Dog's Parsley 143
Dog's Tooth Violet 131
Dogwood 94-5
Doronicum 69
Dorotheanthus 51,55
Douglasia 37
Draba 120-1
Dracaena 84
Dracunculus 130
Dragon Tree 84
Dropwort 70
Druce,G. C. 202
Dryas 121
Dumb Cane 83
Dunham Massey 214
Dunnock 188
Dusty Miller 124
Dutchman's Breeches 68
Dutchman's Pipe 103
Dyer's Greenweed 20
Dyer's Rocket 59
Dyrham 213
Earth Star 80
Earwig 226
Easter Cactus 88
Easter Orchid 81
Eccremocarpus 104

Echeveria 37
Echinacae 69
Echinops 69
Echium 23
Edlweiss 10,122
Edrianthus 121
Egyptian Bastard Rocket 59
Elder 3,99
Eleagnus 95
Elecampane 71
Elephant's Ear 36
Elk's-Horn Fern 87
Embothrium 95
Endive 159
Endymion 18
Epilobium 121
Epimedium 112
Epiphyllum 84
Equestrian Star 85
Equisetum 146
Eranthis 130
Erica 95
Erigeron 28,69
Erinus 121
Erithacus 192
Erodium 70
Eryngium 70
Erysimum 121
Erythronium 130
Escallonia 37
Eschscholzia 37
Eucalyptus 92
Eucharidium 51
Eucharis 77
Eucomis 131
Eucryphia 95
Euonymus 96
Euphorbia 10,14,87,144,150
Evening Primrose 45,52,57,63
Everlasting 119
Everlasting Pea 105
Everlasting Sandflower 47
Exacum 84
Fairy Foxglove 121
False Acacia 16,42
False Aralia 84
False Lily-of-the-Valley 113
False Mallow 75
False Mitrewort 115
Fatshedera 85
Fatsia 84,96
February Fair Maids 131
Felicia 52
Fennel 170
Fennel Flower 57
Fetid Iris 30,131
Feverfew 20,54,67
Ficaria 147-8
Ficus 180
Field Scabious 28
Fig 180
Figwort 24
Filbert 179
Filipendula 6,70
Finger Arelia 84
Firethorn 98

Fittonia 77
Flannel Bush 38
Flame Creeper 106
Flame Flower 79
Flame Nettle 82
Flamingo Flower 79
Flax 54
Flea 226
Fleabane 22,69
Floramor 49
Flower of Bristow 72
Flower of Constantinople 72
Fluellen 29
Fly-Trap 27
Foam Flower 75,115
Foeniculum 170
Fool's Parsley 143
Forget-me-not 56
Forsythia 14,38
Fothergilla 38
Four o'clock Plant 56
Fox 217
Foxglove 45,51
Fragaria 184
Freesia 38
Fremontodendron 38
French Bean 155
Friendship Plant 87
Fringilla 187
Fritillary 26,131
Frog 220
Froghopper 220
Fuchsia 10,33,38
Fumitory 144
Funkia 112
Furze 4
Gaillardia 38
Galanthus 131
Galium 21,27,112
Galtonia 38
Garage 215
Gardenia 38
Garlic 171
Garlic Mustard 30
Garrya 35,104
Gaultheria 112
Gazania 52
Gazebo 214-15
Gean 128
Genista 20,95-6
Gentian 14,35
Geranium 15,70
Gerard, John 200
Gerbera 39
Gesneria 39
Geum 5,70
Gherkin 159
Gilia 121
Gilliflower 68
Gladdon 30,131
Gladiolus 18,131
Glaucidium 122
Globe Amaranth 47,49,52
Globe Flower 75
Globe Thistle 69
Globe Tulip 129

Glory Bower 82
Gloxinia 39
Gnat 226
Goat's Beard 65
Godetia 39,52,58
Gold Dust 118
Golden Rod 75
Golden Trumpet 78
Goldfinch 185
Gomphrena 52
Good Luck Plant 82
Gooseberry 180
Goosefoot 153,155
Goose Grass 27,101
Gorse 4,26,31
Grabowskia 35
Grape 107,180-1
Grape Hyacinth 72,133
Grape Ivy 81
Grass of Parnassus 28,123
Great Tit 195
Greek Valerian 74
Greenfinch 188
Greengage 181
Griselina 39
Grotto 215
Ground Elder 145
Groundsel 28,71,141-2
Guernsey Lily 134
Gunnera 39
Gunpowder Plant 87
Guzmania 80
Gynura 85
Gypsophila 70
Haberlea 122
Habranthus 132,138
Ha-Ha 215
Hairy Bitter-Cress 143
Halesia 39
Hamamelis 96
Harebell 18
Harlequin Flower 136
Hasty Rogers 23
Hawk's Beard 26,120
Hazel 95,178-9
Heart of Jesus 80
Heath 4
Heather 4,94-5
Hebe 96
Hedera 104
Hedgehog 217
Hedge Sparrow 188
Helenium 71
Heleopsis 71
Helianthemum 96
Helianthus 71
Helichrysum 52
Heliophila 52
Heliotrope 27,151
Helipterum 52
Hellebore 30
Hemerocallis 71
Henbane 24
Hens-and-Chickens 125
Heracleum 24
Herb Bennet 5,70

Heuchera 39
Heucherella 112
Hibiscus 96
Hidcote 213
Hillier, Edwin 212
Himalayan Balsam 53
Hippeastrum 128,132
Hoare, Henry 212
Hogweed 24
Holly 20,91
Hollyhock 46-7
Hollyhock Bush 96
Honesty 26,45
Honeysuckle 102,105
Hooker, Sir Joseph 199,202
Hooker, Sir William 202
Hop 102,159
Hop Trefoil 25
Hornet 226
Horn of Plenty 83
Horseman's Star 85
Horse-Radish 2,148,171
Horsetail 146
Hosta 112
Houseleek 125
House Martin 194-5
House Sparrow 189
Houstonia 122
Houttuynia 112
Humulus 159
Hyacinth 18,132
Hydrangea 96,103
Hylomecon 112
Hyoscyamus 24
Hypericum 7,97
Hyssop 171
Iberis 48,112-3
Ice Plant 56
Impatiens 53
Indian Bindweed 104
Indian Cress 9
Indian Mountain Orchid 124
Incarvillea 40
Insect 224 ff
Inula 71
Ionopsidion 122
Ipheion 132
Ipomoea 50,104,162
Iris 132
Irish Ivy 85
Ivy 101-2,104
Ixia 133

Jacinth 18
Jack-by-the-Hedge 30
Jackdaw 190
Jack-in-Prison 57
Jacob's Ladder 25,74
Jade Plant 82
Japanese Bellflower 74
Japanese Knotweed 149
Japonica 94
Jasmine Nightshade 106
Jasminum 104
Jeffersoniana 122
Jekyll, Gertrude 212
Jersey Cabbage 156
Jerusalum Artichoke 154

Jerusalem Cowslip 114
Jerusalem Cross 72
Jerusalem Sage 114
Jew's Mallow 40
John-go-to-bed-at-noon 163
Jonquil 134
Joseph's Coat 47
Jumping Betty 53
Juniper 97
Kaffir Lily 36
Kalanchoe 85
Kale 159
Kalmia 16,40
Kangaroo Vine 81
Kansas Gay Feather 71
Kent, William 211
Kentranthus 53
Kerria 40
Kickxia 29
Kingcup 17
Kingfisher Daisy 52
Knight's Cross 72
Knight's Star 85
Kniphofia 40
Knit-Bone 114
Knotgrass 26
Kochia 53,69
Kohl Rabi 159
Kolkwitzia 40
Lachenalia 85,135
Lacock 213
Lactuca 160
Ladder of Heaven 74
Ladybird 226
Lady-in-the-Bath 68
Lady's Bedstraw 21
Lady's Locket 68
Lady's Mantle 65,110
Lady's Slipper 83
Lamb's Lettuce 21,158
Lamb's Tongues 114
Lamium 113,144
Lamp-Flower 72
Lanhydrock 213
Lapeirousia 133
Lapsane 23-4
Lark's Heels 9
Larkspur 45,68
Lathyrus 104
Laurel 91,98
Laurus 92,169
Lavatera 40
Lavender 97
Lavendula 97
Lawn 216
Layia 53
Leadwort 105
Leaf Cactus 84
Lebanon Candytuft 118
Leek 160
Legousia 59
Leguminosae 95
Lemon 81
Lent-Lily 29,134
Leontopodium 122
Leopard's Bane 69

Lepidum 158
Leptospermum 97
Lettuce 160
Leucojum 133
Levisticum 171
Lewisia 37,40
Liatris 71
Ligularia 71
Lilac 93
Lilium 133
Lily-of-the-Valley 111-12
Lime 20
Limnanthes 37,53
Limonium 53
Linaria 71
Linden 20
Ling 4,94
Linnaea 40
Linum 54
Lion's Mouth 47
Liriope 72
Lithospermum 113
Livelong 5
Livingstone Daisy 55-6
Lloyd, Christopher 212
Lobelia 11,41
Locust 42
Loganberry 181
Loggia 216
Lonicera 35,105
Looper 227
Loosestrife 19,72-3
Lords-and-Ladies 4,5,17
Lovage 64,171
Love Apple 164
Love-in-a-Puzzle 57
Love-in-the-Mist 57
Love Lies Bleeding 47
Lucky Gollans 75
Lunaria 26,54
Lungwort 114
Lupinus 72
Lychnis 46,61,72
Lycopersicon 164
Lyre Flower 68
Lysimachia 19
Lythrum 19,72
Macleaya 73
Madagascar Jasmine 106
Madonna Lily 133
Madwort 118
Magnolia 15,41
Magpie 190-1
Mahonia 41
Maianthemum 29,113
Maidenhair Fern 78,125
Maidenhair Spleenwort 79
Maize 158
Malcolmia 54
Mallow 40,47,54,96
Malope 54
Maltese Cross 72
Malus 176
Mangel Wurzel 155
Manihot 76
Manuka 97

Maori Holly 42
Maple 93
Maranta 85
Marble Plant 80
Marguerite 67
Marigold 48,59
Mariposa Lily 129
Marjoram 69,171
Marrow 160
Marsh Gentian 28
Marsh Marigold 17,18
Martagnon Lily 133
Martin, Rev. Kebel 199
Marvel of Peru 56
Mary's Tapers 131
Mask Flower 47
Masterwort 66
Matricaria 31,54
Matthiola 54
May Lily 29,113
Mayweed 5,54
Mazus 122
Meadowfoam 53
Meadow Rue 75
Meadow Saffron 172
Meadow Saxifrage 125
Meadowsweet 6,70
Meconopsis 27,73
Medicago 25
Medlar 93,181
Melbourne 213
Melissa 168
Melon 181
Mentha 171
Mentzelia 55
Mesembryanthemum 51,55
Mespilus 181
Mexican Aster 50
Mexican Sunflower 60
Michaelmas Daisy 66
Midge 227
Mignonette 59
Milk-Flower 131
Milkweed 66
Milkwort 26
Miller, Philip 199
Millipede 222
Mimosa 31
Mimulus 30
Mint 64,171
Mirabilis 56
Mite 223
Moccasin Flower 83
Mock Orange 93
Mole 219
Moltkia 123
Molucca Balm 56
Moluccela 56
Moly 128
Monarch of the Veldt 60
Monarda 41,169
Monkey Flower 30
Monkshood 64
Monstera 85
Montacute 214
Montbretia 15,41

Moonwort 54
Morisea 123
Morning Glory 50,104
Morus 181
Mosquito 227
Moss 28
Moss Campion 126
Moss Saxifrage 125
Moth 227
Mother-in-law's-Tongue 43,83
Moth Orchid 86
Mountain Laurel 16
Mountain Pansy 126
Mount Atlas Daisy 118
Mouse 218
Mouse-tail Plant 110
Mulberry 181-2
Mullein 29,60,76
Muscari 133
Mushroom 161
Musk 29-30
Mustard 29,162,171-2
Myosotis 56
Myrrh 97
Myrrhis 170
Myrtle 97
Myrtus 97
Nailwort 121
Naked Ladies 129,172
Namaqualand Daisy 60
Narcissus 14,134
Nasturtium 9,106,165
Navelwort 113
Nectarine 182
Nemisia 46,57
Nemophila 57
Neoregelia 80
Nepeta 24,113
Nerine 134
Nettle 143-4
Nettle-leaved Bellflower 28
New Zealand Bur 109
Nicandra 57
Nicotiana 41
Nidularium 80
Nierembergia 123
Nigella 57
Nipplewort 23
Noble Laurel 92
Nolana 57
Nomocharis 135
Nonesuch 72
Normannia 115
Nun's Scourge 47
Nuthatch 191
Nutmeg Flower 57
Oak 20
Obedient Plant 73
Ocimum 169
Oconee Bell 126
Odontoglossum 77
Oenothera 57
Old Man's Beard 103
Oleander 58,98
Olearia 41
Omphalodes 113

Onion 161
Onopordum 58
Ophiopogon 73
Orache 48
Orange 81
Orchid 10,135
Orchid Cactus 84
Oregon Grape 41
Oriental Jacynth 132
Origanum 171
Ormesby Hall 213
Ornithogalum 26,135
Orobanche 26
Orphanidesia 123
Orpine 5
Osier 4
Osmanthus 98
Osmunda 42
Osteospermum 51,58
Oswega Tea 41,169
Ourisia 123
Our Lady' Milk Thistle 59
Oxalis 135
Ox-Eye Daisy 67
Oxford Ragwort 28
Oxlip 28
Ox-tongue 169
Paeony 35,92
Painter's Palette 79
Palm 86
Palma Christi 59
Pancratium 135
Panda Plant 87
Pansy 27
Parahebe 123
Parkinson, John 200
Parnassia 28,123
Parrot's Bill 103
Parsley 172
Parsnip 161
Parterre 216
Parthenocissus 105
Partridge Berry 112
Parus 195
Pasque Flower 29
Passer 189
Passiflora 105
Passion Flower 105
Pastinaca 161
Patio 216
Pea 161
Peace Lily 88
Peach 182
Pear 182
Pearlwort 144
Pearly Everlasting 65
Pelargonium 15,70,103
Penny Royal 171
Penstemon 14,73
Pentaglottis 65
Pepperomia 86
Pergola 216
Perilla 58
Periwinkle 27,115
Pernettya 42
Persian Violet 84

Petasites 151
Petroselinum 172
Petunia 25
Petworth 213
Phacelia 58
Phalaenopsis 86
Phaseolus 155
Philadelphus 93,98
Philodendron 86
Phlomis 114
Phlox 8,73
Phoenix 86
Phylloscopus 196
Physalis 178
Physoplexis 123
Physotegia 73
Pica 190
Piggy-back Plant 43
Pigtail Plant 79
Pieris 98
Pilea 87
Pimpernel 6,119
Pimpinella 168
Pincushion Plant 50
Pineapple 31
Pineapple Lily 131
Pinguicula 123
Pink 68
Piper 86
Pipistrelle 219
Piss-a-Bed 147
Pisum 161
Pittosporum 98
Plantain Lily 112
Plas Newydd 213
Platycerium 87
Platycodon 74
Platystemon 58
Plectanthus 82
Pleione 35,124
Ploughman's Spikenhard 71
Plum 182-3
Plumbago 105
Plume Poppy 73
Poa 142
Poached Egg Flower 53
Poinsettia 87
Polemonium 74
Polesdon Lacey 214
Polianthes 87
Policeman's Helmet 53
Polyanthus 124
Polygala 26
Polygonatum 74
Polygonum 105-6,149
Poor Man's Orchid 59
Pop-Dock 51
Poppy 19,73
Poppy Mallow 120
Portmeirion 214
Portugal Laurel 93
Portulaca 58
Portland Sago 5
Portland Spurge 27
Potato 6,162
Potentilla 98

Poterium 21,74
Poverty 74
Prairie-Button 71
Prairie Lily 55
Prayer Plant 85
Prickly Heath 118
Prickly Phlox 121
Prickly Poppy 48
Prickly Thrift 118
Primula 124
Prince's Feather 47
Prunella 23,114,188
Prunus 98,176-9,187
Pulicaria 22
Pulmonaria 12,114
Pulsatilla 29,136
Purple Rock Cress 36
Pumpkin 162
Puschkinia 42,135-6
Pyracantha 98
Pyretherum 74
Pyrrula 186
Queen's Tears 80
Quince 94,183
Rabbit 217-8
Radish 162
Ragged Robin 52
Rain Marigold 51
Ramonda 124
Rana 220
Ranunculus 136,147-8
Raoulia 125
Raphanus 162
Raspberry 183
Ray, John 201
Rechstieneria 34
Red Hot Cat's-Tail 78
Red Hot Poker 40
Red Ribbons 51
Regal Lily 133
Repton, Humphrey 211
Reseda 59
Rheum 183
Rhipsalidopsis 87
Rhododendron 7,91,98,148
Rhodohypoxis 125
Rhoicissus 88
Rhubarb 64,153,183
Rhus 95,98
Ribes 179
Ricinus 59
Ring Dove 197
Roast Beef Plant 132
Robin 192
Robinia 16,42,92
Robinson, William 212
Rock Cress 119
Rock Forget-me-not 113
Rock Purslane 119
Rockrose 28,96
Rodgersia 75
Romneya 42,48
Romulea 136
Roscoea 42
Rose 9,99
Rosemary 172

Rose-Moss 58
Roseroot 30
Royal Fern 42
Rubus 177,181,183
Rudbeckia 42,69
Rue 172
Rumex 30,164
Runner Bean 102,155
Russian Vine 106
Saffron 48,172
Saffron Spike 79
Sagina 144
St. John's Wort 7,29,97
Saintpaulia 43,88
Salad Burnet 21,23,74
Salix 4
Sallow 4
Salpiglossis 46,59
Salsify 163
Salt Bush 48
Saltram 213
Sambucus 3,99
Sandwort 119
Sanguinaria 125
Sansevieria 43
Santolina 99
Sanvitalia 125
Saponaria 150
Satin Flower 54
Satureia 172
Sauce-Alone 30
Savory 172
Sawfly 227
Saxifrage 9,10,125
Scabious 12
Scallion 163
Schitzanthus 59
Schizostylus 136
Schlumbergera 35,43
Scilla 18,136
Scindapsus 88
Scorpion Grass 56
Scorzonera 163
Scottish Thistle 58
Scrophularia 24
Scurvey Grass 22
Sea Holly 70
Sea Lavender 53
Sea Lily 135
Sea Pink 28,119
Sea Squill 136
Sedum 5,30
Self-Heal 23
Sempervivum 125
Senecio 28,49,71,141
Sequoia 35
Setwall 67
Shallot 6,163
Shamrock 25
Shasta Daisy 67
Shell Flower 56
Shepherd's Purse 26,142
Shillings 54
Shingle Plant 86
Shoo-fly Plant 57
Shooting Star 120

Shortia 126
Shrew 218-9
Shrimp Plant 79
Shrubby Goosefoot 48
Siberian Wallflower 49
Sidalcea 75
Silene 61,126
Silk Tassel Bush 104
Silver Vine 88
Silverweed 25
Silybum 59
Sinapsis 171
Sinningia 39,43
Sisyrinchium 136
Sita 191
Sizergh Castle 213
Skimmia 91,99
Slipper Flower 119-20
Slug 221
Small Medic 25
Smithianthus 43
Smith, Sir James Edward 199
Smoke Bush 95
Snail 221
Snake's Head 26
Snapdragon 47
Sneezewort 71
Snowbell 126
Snowberry 149-50
Snowdrop 131
Snowdrop Tree 39
Snow-in-Summer 111
Snowy Mespilus 93
Soapwort 150
Solanum 36,106,162
Soldanella 126
Solidago 75
Solomon's Seal 74
Sonchus 24
Song Thrush 193
Sops-in-Wine 68
Sorbus 99
Sorrel 30,163
Sow Bread 130
Sowerby, James 202
Sow Thistle 23
Spanish Bayonet 76
Spanish Violet 72
Sparaxis 136
Sparrow 188-9
Spathiphyllum 88
Spectacle Plant 54
Specularia 59
Speedwell 29,56,149
Spider 222-3
Spider Flower 49,81
Spiderwort 25,44
Spinach 3,6,164
Spindle 6,96
Spiraea 70,99
Spleenwort 79
Split-leaved Philodendron 86
Spotted Dog 114
Spotted Medic 25
Sprekelia 137
Spring Snowflake 133

Spring Star Flower 132
Spur-flower 82
Spurge 6,144,150
Spurge Laurel 92
Squill 29
Squinancywort 111
Squirrel 219
Stachys 114
Stag's-Horn Fern 87
Star Grass 125
Starling 193-4
Star of Bethlehem 25,135
Starry Jacinth 136
Starwort 66
Stellaria 142
Stephanotis 106
Sternbergia 43
Sternus 193-4
Stinking Iris 131-2
Stinking Camomile 6
Stitchwort 142
Stock 54-5
Stock Gilliflower 49,55
Stone Cress 118
Stourhead 212
Strawberry 64,148
Strawberry Tree 95
Straw Flower 52
Streptocarpus 88
Streptopelia 187
Strelitzia 43
Stripped Squill 42
Stuartia 43
Succory 157
Sumach 98
Summer Cypress 53
Summer Hyacinth 38
Summer Snowflake 133
Sundew 17,26
Sunflower 27,45,71
Sun Plant 58
Swallow 194-5
Swallow-Wort 29
Swan River Daisy 48
Swede 164
Sweet Cicely 30,170
Sweet Flag 21
Sweet Pea 102,105
Sweet Potato 162
Sweet Vernal Grass 21
Sweet William 49,68,119
Swine's Snout 136
Swiss Cheese Plant 86
Sylvia 185
Symphoricarpus 149-50
Symphytum 114
Syringia 93
Tagetes 17,46,59
Tamarisk 99
Tansley, Sir A. G. 199
Tansy 67
Taraxacum 146
Tare 27
Tarragon 173
Tasetta 134
Tassel Flower 47

Tassel Hyacinth 133
Tea Tree 97
Tecophilaea 137
Temple Bells 43
Temple Newsam 214
Terrace 216
Thalictrum 75
Thimble Flower 51,121
Thistle 146
Thorn Apple 83
Threepenny-Bit-wort 119
Thrift 119
Thrips 227
Thrower, Percy 212
Thrush 193
Thunbergia 35,106
Thuya 99
Thyme 26,173
Thymelaecae 92
Thyme-Olive 92
Tiarella 115
Tickseed 67
Tidy Tips 53
Tiger Lily 133
Tigridia 137
Tit 195
Tithonia 60
Toadflax 29,71
Tobacco 41
Tolmiea 43
Tomato 36,164
Topiary 216
Torch Lily 40
Torenia 44
Touch-me-not 53
Townhall Clock 25
Trachelospermum 106
Trachymene 60
Trachystemon 115
Tradescantia 25,44,88
Tragopogon 163
Transvaal Daisy 39
Treacle Mustard 121
Treasure Flower 52
Tree Celandine 73
Tree Poppy 42
Trellis 216
Trillium 115
Tritonia 137
Troglodytes 197-8
Trollius 75
Tropaeolum 9,106
Tropical Bleeding Heart 82
Trumpet Creeper 103
Tuberose 87
Tulip 10,137-8
Turdus 186
Turk's Cap 133
Turner, William 200
Turnip 165
Tutin, T. G. 203
Tutsan 7
Twayblade 25
Twin Flower 40
Twin Leaf 122
Twin Spur 50

Umbelliferae 168
Umbrella Grass 83
Ursina 60
Urtica 143
Vaccinium 177
Valerian 49,67
Valerianella 21,158
Vallota 138
Vegetable Oyster 163
Venidium 60
Venus' Looking Glass 59
Verbascum 60,75
Verbena 45,60-1
Veronica 96,149
Vervain 60
Vetch 27
Viburnum 100
Vicia 154
Vinca 115
Viola 10,126
Violet Cress 122
Viper's Bugloss 23,65
Virginia Creeper 101,105
Virginian Stock 54
Viscaria 61
Vitis 107,180
Vole 218
Vriesia 80
Wake-Robin 5
Waldsteinia 115
Walking-stick Cabbage 45
Wallflower 27,45,49,63,68
Walnut 27,184
Wandering Jew 44,89
Wand Flower 130
Warburg, E. F. 203
Wasp 227
Watercress 24,165
Waterlily 15
Wayfaring Tree 100
Wax Flower 106
Weevil 227-8
Weigela 44
Weld 59
Welsh Onion 27
Welsh Poppy 27,73
Whin 4
White Pipe-Tree 93
Whitlow Grass 121
Whortleberry 177
Wig Tree 95
Wild Ginger 110
Willow 4,12,20
Willowherb 58,121
Willow Warbler 196
Winter Aconite 29,130
Winter Daffodil 43
Winter Heliotrope 151
Wishbone Flower 44
Wisley 213
Wisteria 35,107
Witch Hazel 96
Withy 4
Wolf Pea 72
Wolf's Bane 69
Wood Avens 5

Wood Pigeon 197
Woodlouse 222
Woodruff 21,111
Wood Sorrel 25,28,135
Worm 220-2
Wormwood 22-3
Woundwort 23,114
Wren 197-8
Wych Elm 96
Yarrow 26
Yellow Archangel 113
Yew 91
Young, Andrew 199
Youth-and-Old-Age 44
Youthwort 17
Yucca 76
Zantedeschia 35,44
Zea 158
Zebra Plant 79
Zebrina 88
Zenobia 100
Zephyranthes 138
Zinnia 44
Zucchini 160
Zumbuls 132